A SAGA OF SEA EAGLES

John A. Love

Whittles Publishing

Published by
Whittles Publishing Ltd.,
Dunbeath,
Caithness, KW6 6EG,
Scotland, UK

www.whittlespublishing.com

ISBN 978-184995-080-0

Also by John Love:

Eagles (1989)
Sea Otters (1990)
The Return of the Sea Eagle (1993)
Penguins (1994, 1997)
A Salmon for the Schoolhouse (1994)
Rum: a landscape without figures (2001)
A Natural History of St Kilda (2009)
The Island Lighthouses of Scotland (2011)

Printed by Charlesworth Press, Wakefield

Contents

Preface v

1 The return of the native 1

2 Sea Eagle facts 12

3 Sea Eagle fiction 28

4 Food and feeding 44

5 Sea Eagles in Britain 59

6 Sea Eagle persecution 82

7 Reintroduction 99

8 Rum 121

9 Adapting to the wild 140

10 The native has returned 161

11 Further developments 183

12 Living wth Sea Eagles 214

Postscript 233

References and further reading 234

Appendix I White-tailed Sea Eagle in different languages 240

Appendix II Current world status of White-tailed Sea Eagles 242

Appendix III Numbers of Norwegian Sea Eagles released in the UK and Ireland 244

Appendix IV Establishment of the Sea Eagle population in Scotland 245

Index 246

For my parents Howell Clifford Love (1910–1955)
and Agnes Grant Fraser Watson (1918–2009),
my elder brother James (Jim) Love (1943–2006),
and of course, David, Maureen, Heather, Jamie and Lucy

Preface

There be...things which are too wonderful for me...
The way of an eagle in the air....
Proverbs 30:18–19

In 1998 I was passing through London on my way to Kenya for a holiday (hoping, as it happened, to see African Fish Eagles, amongst other things). I stopped off at the famous Foyles bookshop, then the largest in the world, to stock up on some holiday reading from the natural history section, and paid by cheque. Examining my signature, the sales assistant, who was probably in her late thirties, inquired: 'Are you the John Love who wrote *The Return of the Sea Eagle*?' I have never considered myself a famous author and had never had any hopes that my first book would be a bestseller. What's more, the book had been published back in 1983, when the assistant must have been just starting out on her career, and it had been out of print for several years. With staff like that, no wonder Foyles was famous.

I think only 2,000 or so copies were printed and the book has long attracted handsome prices second-hand. Over the years, many people have asked for it to be republished and even more have wished to see it updated. I had chosen a monographical approach to *The Return of the Sea Eagle* since, at the time, the species was still relatively unknown. What research had been done was mostly to be found in inaccessible or foreign journals.

Sea Eagles had not nested in Britain for 70 years or more and, since 1975, I had been managing a pioneering reintroduction project in the Hebrides. My first intention in writing the book had been to present a background to the species and its biology. I felt it was important to take a historical approach to justify – for the first time – why it should be reintroduced. This was, at the time, a fairly novel slant: the only other book I knew like this had been George Waterston's classic *The Return of the Osprey*, written in 1962. It is no coincidence that I adopted a similar title. Indeed, I dedicated the book to George, and to Dr Johan Willgohs of Bergen University in Norway, who had helped George in a pioneering reintroduction attempt on Fair Isle. Both became good friends of mine but, sadly, are now no longer with us. The third dedication was to one of my closest friends, Harald Misund of Bodø in northern Norway, with whom I am in constant touch to this day, and without whom the Rum reintroduction would never had happened. British ornithology and conservation owe much to Harald.

The Return of the Sea Eagle was published in 1983, perhaps a little prematurely, for we had not yet stopped importing young Sea Eagles from Norway, while the embryonic wild population in the Hebrides had only just laid its first egg. It would be another two seasons before the first Gaelic-speaking Sea Eagle fledged in Scotland and, as we shall see, a lot has happened since then.

There is clearly a need for an update. This book, though, will not be *The Return of the Sea Eagle* with a few new chapters added. Over the years there have been so many people involved in the Sea Eagles' story, forging significant careers in conservation as a result (so where did I go wrong?) that I feel it inappropriate now to tread ground that is less familiar to me. There has been much research undertaken in recent years, much of which is summarised in a volume of papers delivered at a highly successful symposium in the year 2000 in Björkö, organised (and edited, along with Mick Marquiss and Bill Bowerman) by my friend Björn

Helander, champion of Sea Eagles in his native Sweden. It is likely that another international conference will be convened soon, hopefully in Scotland, to bring the situation bang up-to-date.

Consequently, I have called this book, which is more of a personal memoir, *A Saga of Sea Eagles*. The Rum phase ended in 1985 but I continued to help with monitoring progress in the field, and remain to this day one of only two independent members of the UK Reintroduction project team, the other being Roy Dennis. There have now been two other phases in the Scottish reintroduction, in Wester Ross and in East Scotland, while a separate reintroduction is taking place in Ireland, with proposals mooted, too, for England and Wales. Although I shall summarise these for the sake of completeness, it is better that they be documented by those most intimately involved with them. I should, however, like to acknowledge their efforts, along with the countless people along the way, sadly now too numerous to mention individually, who have helped re-establish the White-tailed Sea Eagle in its old breeding haunts.

I have, in places, dipped heavily into my previous book, mostly on the historical side, since the approach I took then still appears to be an unusual one, while a lot of the material, previously submerged in numerous old and obscure texts, still merits being recycled in a coherent manner. There is, after all, a whole new generation, maybe even two, who will be unfamiliar not only with my first book but also with the background to the project. Not everyone can be expected to have the encyclopaedic memory of a diligent assistant in Foyles bookshop!

And now for some explanation of terminology. First of all I promote, as I always have, the spelling of 'Rum', as it should be, whether in Gaelic or Norse, rather than 'Rhum'. I also adhere to *Haliæetus albicilla* as the Sea Eagle (with capital letters), reserving 'sea eagle' in lower case for a member of the genus *Haliæetus*. Although this has been the convention ever since Morton Boyd, Pat Sandeman, George Waterston and Roy Dennis first initiated reintroduction attempts in the UK, it now appears to be old-fashioned and out of date. Nowadays we are required to use the term 'White-tailed Eagle'. As will become clear in the next chapter, 'White-tailed Eagle' implies that the species is something it is not. It is not a true eagle in the strict taxonomic sense but a rather distantly related sea eagle.

My very first bird guide was *The Observer's Book of British Birds*, published in a revised edition in 1952 and bought by my parents, for the princely sum of five shillings, for my Christmas two years later. Not only did it opt still to include the extinct *Haliæetus albicilla* but quite happily used both names – Sea Eagle and White-tailed Eagle. More recently, the authors of *Raptors of the World* (2001), perhaps rather unhelpfully, have elected to refer to the entire genus *Haliæetus* (except, for some reason, the Bald Eagle) as 'fish-eagles'. So, in this hefty and authoritative tome, *Haliæetus albicilla* is called the White-tailed Fish-eagle! It's true, all members of the genus eat fish to a greater or lesser extent and true, too, that not all *Haliæetus* are entirely restricted to the sea. To distinguish this distinctive genus as 'sea eagles', though, still seems to be entirely reasonable.

I lament this recent standardisation of English names – was it just for the convenience of our cousins across the Pond? Sadly, we risk losing all our glorious regional names. I delight, for instance, in lapwing, peewit, peesie, green plover, teuchit, hornpie, flopwing, cornchwiglen, curacag or whatever, rather than Northern lapwing. If a foreigner cannot understand, then, for that very reason, the scientific name *Vanellus vanellus* is always there. I had always believed such Latin binomials, rather than English, to be the *lingua franca* of taxonomy. Jeremy Mynott, in his *Birdscapes* (2009) offered an entertaining and informative discourse on this and many other subjects. Call me old-fashioned if you like, but I will persist in referring to the subject of this book as the Sea Eagle or, occasionally, the slightly more cumbersome White-tailed Sea Eagle, and its congeners as sea eagles.

What one actually calls the Sea Eagle will, of course, depend upon which country you come from, or even which part of that country you come from. Every language has its own name for a species and the Sea Eagle is no exception (see Appendix I). Just within Scandinavia, for example, the Sea Eagle is 'Havørn' in Norwegian, Danish and Faroese, 'Havsörn' in Swedish and 'Haförn' in Icelandic. All are obviously from a common root and give us the Old English term Erne for eagle. In the Celtic languages, the Sea Eagle is 'Er' in Cornish and Breton, 'Eryr cynffonwyn' in Welsh, 'Iolair mara' in Irish and 'Iolaire mhara' in Scots Gaelic. Gaelic can treat the word 'Iolaire' as either masculine or feminine and I will stick to the latter, necessitating the aspiration of the following adjective.

Gaelic has a variety of other names too: 'Iolaire bhann' or 'Iolaire fhionn', meaning the pale or white eagle, 'Iolaire ghlas' – the grey eagle – or 'Iolaire chladaich', the shore eagle. The distinctive speckled-brown plumage of the juvenile merits the appropriate 'Iolaire bhreac' or 'Iolaire riabhach' (an 'r' is not aspirated). Best of all, there is also an eloquent poetic term for the adult: 'Iolaire sùil na grèine', the eagle with the sunlit eye. It is the Golden Eagle that is accorded the distinction of 'Fir eun', the true bird. The word 'iolaire' is pronounced a bit like 'yulir', deriving from the related Celtic languages of the ancient Britons and Irish, and in some place names in the Lake District it seems to have been corrupted to 'wallow'. The thick Lakeland brogue sometimes also mutilates the old word 'erne' to 'iron' or even 'heron', so somewhere like Heron Crag in Cumbria may be an ancient eagle site and nothing to do with herons at all.

The use of the word 'Erne' for the Sea Eagle is from an Old Norse or Anglo Saxon word, 'ørn', meaning 'the soarer'. This probably has an ancient root in the Greek word 'ornis' for bird, the word, indeed, from which 'ornithology' derives, making an etymological link between the science itself and eagles. 'Ørn' survives throughout Britain in old place names, such as Arnecliffe in the Yorkshire Dales and Ernesheugh, inland from Dundee, with another on the coast just south of Aberdeen. The name is especially common in Shetland and Orkney, with Erne's Brae, Erne's Hamar, Erne's Heugh and Earn Stack.

The spelling 'earn' has to be treated with caution. In many instances it does indeed refer to the Sea Eagle, but while we have seen how it can be corrupted to Iron or even Heron Crag, the ancient word 'earn' might also mean white, as in the river and loch of that name in Perthshire. It is also used as a person's name, so may not imply an old eagle site but merely 'Arn's cliff' or 'the cliff of Arn'. And here's another word of caution about the much more familiar word 'eagle'. Places such as Eaglehall or Eaglesham are more likely to derive from the Celtic word 'eaglais', meaning 'church'. It's true, though, that Eaglescliffe on the Yorkshire coast would have provided a rather precipitous place of worship and so it, at least, may really have been a haunt of Sea Eagles. In the same county are found Arncliffe, Arnecliffe and Yarncliffe.

It is thought that place names such as Arley in Cheshire, Earley in Berkshire and Earnley in Sussex derive from 'earn-leah' meaning 'eagle wood', and Earnwood in Shropshire and Arnewood in Hampshire may have a similar derivation. It may be no coincidence that many of these names are to be found close to major river systems such as the Severn and the Thames while, of course, the Somerset Levels, the Fens, the Broads and so on would all have offered prime Sea Eagle habitat. No longer breeding in the south of Britain, and latterly still well known as winter visitors, these Sea Eagles – mostly immatures – were often called Fen Eagles.

The distinguished 18th-century naturalist Thomas Pennant reserved Erne, along with Cinereous Eagle, for the adult of the species, while he classified the distinctive immature separately under the term Sea Eagle. The famous 19th-century American bird artist John James Audubon fell into the same trap, referring to the young Bald Eagle as the Washington Eagle. The Americans call our British species the Gray Sea Eagle, while some old English texts refer to it as the Fish-eagle, immediately inviting confusion, not only with fish eagles elsewhere in the world but also with the archetypal 'Fish Eagle' or 'Fish Hawk', the Osprey. To add to the muddle, indeed, Aristotle, in his *Natural History* written over 2,000 years ago, called the Osprey the 'Sea-Eagle'.

Taxonomists have failed to agree even over its scientific name. In 1758, Linnaeus included the Sea Eagle in the genus *Falco* and it was not until 1809 that the French zoologist Marie-Jules César Lelorgne de Savigny (1777–1851) was to create the modern genus *Haliæetus* for sea/fish eagles. The word is sometimes spelt '*haliaeetus*' or even '*haliaetus*', but should properly be rendered *Hali-æ-e-tus*. Thus, under the Linnaean system as modified by Savigny, our Sea Eagle species became *Haliæetus albicilla*. *Haliæetus* derives from the Greek 'hali', for 'salty' or 'of the sea', while 'aetos' means eagle. This, incidentally, is also the scientific name given by Savigny to the Osprey *Pandion haliæetus*. The specific name for the White-tail is *albicilla* and means exactly that in Latin. One can further reduce this to subspecies level by the addition of a third, subspecific name *Haliæetus albicilla albicilla* or, in short, *H.a.albicilla*. This refers to the population widespread across Europe and Asia, while the slightly larger *H.a.groenlandicus* is confined to southwest Greenland. Following Linnaeus, incidentally, the specific and subspecific names are always in lower case, while only the genus is accorded a capital letter.

There is a poetic collective noun for eagles – a convocation – though quite where this term came from is obscure. With an uninspiring dictionary definition of 'an assemblage or gathering called together for a purpose (and often used by the clergy)', might I suggest a more palatable alternative for the Erne – a saga of sea eagles?

And one final comment. When I use the term 'sea eagle' I have to be precise in my pronounciation for, occasionally, I have been confronted with a perplexed response: 'Why on earth did you want to reintroduce seagulls?'

Chapter 1
The return of the native

What a joy it would be if we could get the Sea Eagle back to Scotland as a breeding species...
George Waterston, 1964

Monday, 24 June 1968

> The wind was quite strong with heavy rain. We remained at the Observatory in the morning and had a slow walk south after lunch. We were overtaken by Roy [Dennis] and company, driving up to the airstrip to meet – we were told – 'the Sea Eagles', so naturally we followed. The RSPB had chartered a Loganair Islander to fly in three seven-week-old eaglets from Bergen in Norway. The Norwegian Sea Eagle expert Dr Johan Willgohs and his wife were aboard and accompanied Roy and George [Waterston] and saw the birds safely installed into their cage on Erne's Brae. [A fourth was to arrive a few days after we left]. They will be released when they are old enough to fend for themselves, and it is hoped that in future years they will breed here in Scotland. Andy Allsop, the pilot, took off again, almost sideways in the strong wind! . . . We were up at 0500 next morning and had a very calm and sunny crossing to Shetland in the island mailboat *Good Shepherd*.

This is an extract from my diary when, with fellow student Bruce Philp, I was visiting Fair Isle on holiday. (I was to meet up again with Andy Allsop, the Loganair pilot, twenty two years later in the Falklands, where he was training pilots for FIGAS, the Falkland Islands Government Air Service.) Roy Dennis, the Bird Observatory warden, had been a friend of mine from my schooldays. In 1963–1964, taking time off school, I had volunteered for Operation Osprey at Loch Garten, when Roy had been warden there. This venture had been a bold move to let the public enjoy the first pioneering pair of Ospreys in Speyside, and had been masterminded by George Waterston, then Scottish Director of the RSPB. In another bold venture, back in 1948, George had also founded the Bird Observatory on Fair Isle.

We had already suspected that something big was about to happen on the island when, a few days earlier, George Waterston had turned up at the Observatory accompanied by Shetlanders Theo Kay and Denis Coutts and John Arnott of the BBC. That same evening we had been privileged to sit with them around the stove, listening to old Theo recounting fascinating tales of Shetland. I often wonder if John recorded them.

In 1959 George had been involved with an effort by his cousin Pat Sandeman to release

The first Sea Eagle arrives at Fair Isle by Loganair Islander in June 1968, with Johan Willgohs and George Waterston, Roy Dennis and pilot Andy Alsopp behind (Denis Coutts)

three young Sea Eagles in Glen Etive, Argyll. Sadly, neither these – nor the Fair Isle Sea Eagles – went on to breed, too few in number to establish any sort of viable population. Little did I suspect then, however, that only seven years later I myself would be intimately involved in a third reintroduction attempt – an ultimately successful one – on the Isle of Rum in the Hebrides. Roy was involved in the planning of this project with the Nature Conservancy Council and recommended me for the short-term contract. This was eventually to develop into a career . . .

For me, at least, it all began on that bleak, windswept moor on Fair Isle. The idea, though, had been hatched some time earlier by Pat, George and a rookie Regional Officer with the Nature Conservancy, John Morton Boyd. A few years ago, while I worked as Area Officer covering Uist, Barra and St. Kilda for Scottish Natural Heritage (a successor body to the Nature Conservancy and subsequently the Nature Conservancy Council) I spent a little time perusing old files in search of background about our local National Nature Reserves. Not really in any mindset for Sea Eagles, I was somewhat surprised to come across some intriguing correspondence. It is here, really, at remote St. Kilda, that our saga of reintroduced Sea Eagles begins.

In December 1956, the archipelago of St. Kilda had passed into the ownership of the National Trust for Scotland. Having no money to run the property, they were to rely heavily on the Nature Conservancy, who declared it a National Nature Reserve. The islands, along with the rest of the Hebrides, would come under the remit of a new Conservation Officer based in Edinburgh, John Morton Boyd (or Morton Boyd, as he liked to be known).

Born in Darvel, Ayrshire, in 1925, Morton Boyd did his National Service in the RAF before studying zoology at Glasgow University. It was with a group of fellow students that he first visited St. Kilda, in July 1952, the year before he graduated with an Honours degree at the age of 28. He went on to undertake a PhD on earthworms in Tiree but was able to return to St. Kilda in 1955 and again in 1956. In July 1957 he was appointed to the Nature Conservancy as Conservation Officer for the northwest of Scotland, after which, under his new remit, his visits to St. Kilda became routine. He would later become a key figure in Scottish conservation and, ultimately, Scottish Director of NCC (1971–1985). He was always a keen supporter of the Rum project and I now realise why.

I am not sure what may have first inspired Morton but on 8 March 1957 – a few months before he began his career with the Nature Conservancy – he wrote to its Director General in London, the eminent conservationist Max (later Sir Max) Nicholson:

Morton Boyd at the eagle cages in Rum, July 1976

> In principle I am strongly opposed to man-made introductions into natural ecosystems, and particularly so at St. Kilda. Yesterday, however, at the National Trust in Edinburgh I made the suggestion that the White-tailed Eagle might be reintroduced at St. Kilda... I have often thought, while on Conachair, the Cambir, or Boreray how perfectly this great bird would fit into the complex, given absolute protection...

Unfortunately I knew nothing of this letter when Morton used to visit me, first on Rum and latterly, after his retirement, in South Uist. He died in 1998 so I never had the chance to discuss it, nor did Morton mention his early exchange about Sea Eagles in his autobiography, published posthumously a year later. I know he was passionate about St. Kilda and about Sea Eagles. Ever the romantic, I can easily picture Morton sitting on Conachair, the highest sheer cliff in Britain, pondering their return.

If his was not the original idea, it could be that he had already spoken with Pat Sandeman – whom he refered to as 'an ornithologist well known for his work in Scotland in the cause of bird protection, his especial interest being birds of prey' – who may then have sown the first seed in his mind. We may never know, but Morton certainly went on to encourage Pat formally to write to the Nature Conservancy on 31 July 1957:

> I put forward a formal request that, in the event of my being able to obtain a pair of White-tailed Eagles (*Haliæetus albicilla*) from Norway, where the bird is heavily persecuted, you will accept the pair for release at St. Kilda.
>
> As you may be aware, the Norwegian Government and the Gamekeepers' Association of Norway both pay substantial bounties for White-tailed Eagles which are shot or captured. It is through a correspondent who is interested in the birds that I may obtain a pair that has been trapped.
>
> The White-tailed Eagle is an indigenous species exterminated by persecution early this century and it would be very interesting to try and establish a pair in an outlying area like St. Kilda, in view of the dwindling numbers in other parts of Europe.

Pat was born in 1913, into the Sandeman Scotch whisky and port family. Brought up in Edinburgh, he was commissioned into the Royal Artillery to serve much of World War II on the front line. He was deeply interested in Highland culture, learning to play the bagpipes from the famous Pipe Major Willie Ross and, always a keen ornithologist, he regularly tramped the hills and glens surveying golden eagles. He was immediately attracted to the idea of reintroducing White-tailed Sea Eagles to Scotland and Morton readily registered his support to the Nature Conservancy. Pat's letter seems to have passed to the next stage for, at a

Pat Sandeman (Mari Sandeman)

meeting held in Hope Terrace (the Nature Conservancy Scottish headquarters) on 23 January 1958, the Nature Conservancy formally agreed to the proposal. But it never came to fruition. I am not aware of any reference to Sea Eagles in old National Trust for Scotland files, though one probably exists somewhere, but I suspect the NTS disapproved of the idea.

Pat Sandeman had to look elsewhere. He chose Glen Etive in Argyll, as explained in his note in *Scottish Birds* (1965):

> In July 1959 an adult and two young of that year were ransomed from Norway, where the species is much persecuted, and were loosed in Glen Etive by kind permission of the proprietors. They were kept in the vicinity of a wood for a fortnight, tethered on a long cord near a pool and within reach of the carcass of a dead hind. They were then released.
>
> Unfortunately the adult was semi-tame and appeared by the roadside, where people photographed it. Not surprisingly it was captured about a month later by an Appin farmer, whose hens it was attacking. After travelling to the London Zoo as a Golden Eagle it was eventually returned, properly identified, to live in the Edinburgh Zoo.
>
> The two young birds, however, were very wild and learned to fend for themselves, being seen several months later soaring in the hills. They were heard calling in flight and clearly identified by their wedge-shaped tails. One was later found in a fox trap some 50 miles south at Otter Ferry in January 1960.

The redoubtable climber Tom Weir regularly met up with Pat and many years later gained some additional detail which Weir published in his regular feature 'My Month' in the *Scots* magazine (1982). He told Tom how his enthusiasm for eagles went back to his boyhood with his cousin George Waterston. Concerned at the high level of persecution, George and he were later instrumental in establishing a bounty system to reward gamekeepers for fostering successful breeding in their local eagle pairs rather than selling the eggs to collectors. Pat explained about his Sea Eagles:

> Did I ever tell you about 'Norway's Golden Gift to Scotland'? That was a newspaper headline when a Norwegian called Reidar Brodtkorb got in touch with George and me, offering us a golden eagle to be set loose on our hills to save it from persecution by Norwegian farmers. We gave it its freedom in the Trossachs.
>
> The next thing was a letter from Trondheim offering us three white-tailed sea eagles. Two were in a nest that was about to be destroyed and the other was in a cage. We consulted Captain Bell of Glen Etive and asked for his help. 'Bring them here,' he said. At Waverley Station we found three screaming white-tailed eagles and a heavy pong of guano in the air. The police were there and curious sightseers milling around. As quickly as we could, we rushed the birds off to the foot of Glen Etive.
>
> We put the eagles in a cellar for the night with venison and water, and first thing in the morning we climbed to a suitable place in the forest, where there were pools of water,

and where two dead hinds had been placed. Seton Gordon, the naturalist, was there, with Captain Bell and his family. Whenever we let go the birds they jumped into the water and wallowed in the pools, loving it after captivity. The birds had ten yards of pinion around their legs which kept them tethered for three weeks until we reckoned they could fend for themselves. Then we waited for reports of sightings.

The one that had been caged was a publicity seeker. It settled at the roadside in Glencoe and was seen being photographed by tourists. Two weeks later it was caught in a net in Appin trying to get into a chicken run. It was taken to London Zoo, then sent to Edinburgh Zoo where it died in 1963.

As for the two birds that had been taken as chicks from the nest, one was seen in company with a golden eagle flying over Glen Etive. A year after its release, one was caught in a trap near Otter Ferry on Loch Fyne.

The operation was kept secret because of the prejudice against eagles at that time. Pat referred to the third bird as an adult but from his photograph which accompanied Tom Weir's article it appears to be no more than two or three years old. I corresponded with Pat before he died, and it is a pity that no other account survives of this pioneering reintroduction. Indeed he had only been inspired to submit his brief account to Scottish Birds in 1965 by two papers by his cousin. The first article had appeared a year earlier in the RSPB's Bird Notes, describing George and Irene Waterston's week spent with Johan Willgohs in his motor launch Sterna around the Bergen area in western Norway.

Johan Willgohs in Norway 1964 (George Waterston)

> Our quest was the Sea Eagle, and in Johan we had an expert guide. He has spent many years studying this species and had published a valuable paper…What a joy it would be if we could get the Sea Eagle back to Scotland as a breeding species. At one time they were far more common than the Osprey in the Highlands – and in fact more common than the Golden Eagle…The Osprey has now returned; let us hope that one day the Sea Eagle may do likewise.

George contributed to a second article in British Birds (1964) accompanying a selection of black-and-white prints of Sea Eagles taken by various Scandinavian photographers:

> Following the successful re-establishment of the Osprey *Pandion haliæetus* as a breeding species in Scotland, there has been a certain amount of speculation on the possibilities of the return of the White-tailed Eagle *Haliæetus albicilla* to one of its ancestral haunts. . . Although there was an unsuccessful attempt at an artificial reintroduction of a couple of young birds from Norway in 1959, and unsubstantial rumours of others nesting, the species today is but a rare vagrant in the British Isles.

George Waterston was a hugely influential figure in Scottish ornithology and few who met him, myself included, failed to respond to his dignified but kindly presence, his vision, determination, knowledge and enthusiasm. His visits to the Inverness Bird Club were a great inspiration to me as a schoolboy, and in the early 1960s had encouraged me to apply as a volunteer for Operation Osprey in Speyside. This fish-eating hawk had also been hounded to extinction in Britain early in the 20th century. Being regular migrants to West African wintering grounds, a few Ospreys would stray off course into the Highlands. When a pair began to breed there in the late 1950s, George soon opened up a hide to the public. Two years later, in 1960, he recruited Roy Dennis, whom he had met when Roy was an assistant warden on Fair Isle. Roy recounts the full story of Operation Osprey in his wonderful book *A Life of Ospreys* (Whittles, 2008).

George Waterston (1911–1980) in Fair Isle in 1969 (Denis Coutts)

George himself had been greatly influenced by Evelyn Baxter and Leonora Rintoul, those grand old ladies of Scottish ornithology whose two-volume *Birds of Scotland* (1953) quickly became a classic and gave the first comprehensive accounts of the demise of both the Sea Eagle and Osprey. George followed the ladies to the Isle of May in the Firth of Forth, with its embryonic bird observatory, before making his first visit to Fair Isle in 1935. Appreciating its own potential as a bird observatory George enlisted Theo Kay and his friends to assemble the first Heligoland traps. The outbreak of war took George to serve with the Royal Artillery in Crete which, in 1941, fell to the Germans. After two years in a prisoner-of-war camp (with, as it happened, another Scottish ornithologist, Ian Pitman,) he had to be repatriated with kidney problems from which he never fully recovered. In the camp, though, he had occupied himself with devising detailed plans for the bird observatory on Fair Isle, and indeed this island was his first sight of home on the voyage back to Britain. After the war, in 1948, George was able to purchase Fair Isle for £3,500 before passing it to the National Trust for Scotland in 1954.

Despite all his other commitments, George never gave up on his vision of seeing Sea Eagles fly over Scotland once more. In a feature in the RSPB's *Birds* magazine (1968) he explained:

Fair Isle Bird Observatory in 1966

> Over 100 years ago, the White-tailed or Sea Eagle (or Erne, as it was usually called in the north) was more common in Scotland than the Golden Eagle. Sheep farmers were always its traditional enemies, and its habit of feeding on sheep carrion was enough to seal its fate. As a result of shooting, trapping and poisoning, numbers were reduced to a few pairs and the coup de grâce was delivered by egg and skin collectors. The last pair in Shetland was robbed in 1910 and did not nest again. In 1916, the last pair nested in Skye; man had finally exterminated the Sea Eagle as a breeding species in Scotland.
>
> Sea Eagles are mainly sedentary, and unlike the Ospreys which have recently recolonized in Scotland as a result of immigration from Scandinavia, it is considered highly unlikely that they would ever return to their former haunts unaided. As Man was responsible in the main for the extermination of this species, the RSPB decided to make an effort to reinstate it: hence the experiment to introduce some young Sea Eagles on Fair Isle.
>
> After many years of close study, Dr Johan Willgohs, the noted Norwegian authority on the Sea Eagle in his country, has shown conclusively that the bird preys primarily on seabirds and fish; it is also a carrion-eater and scavenger. He has never, in the course of many years of field work, seen one attack a lamb. Indeed, he has watched a sheep with a lamb grazing unconcernedly below an occupied eyrie.
>
> Fair Isle was the location chosen for the experiment; the Sea Eagle last nested there in 1840. It has many advantages: it is one of the most remote inhabited islands in Britain, being 25 miles from Shetland, its nearest neighbour; and it has an abundance of prey species – seabirds, fish and rabbits. By no means the least important aspect is that this project has the unanimous support of the islanders. The National Trust for Scotland, owners of Fair Isle, gave the experiment their blessing, and approval was given by the Nature Conservancy and the leading ornithological organisations.

Young Sea Eagle in its cage in Fair Isle, June 1968 (Norman Elkins)

It was decided to bring four young birds from Norway to Fair Isle, and Dr Willgohs agreed to try to procure two males and two females, all from different eyries in north Norway where they are habitually destroyed. Permission to collect and export these birds was granted by the Norwegian Government.

After a preliminary reconnaissance in spring, Dr and Mrs Willgohs returned to the north in mid-June and, after much toil and trouble, collected three young birds which were flown with them from Bodø to Bergen. On 24 June a small Loganair charter plane flew the party into Fair Isle via Kirkwall. The birds were installed in big roomy cages and were soon having a good feed of fish and rabbits. The fourth youngster, which had arrived too late at Bodø, was eventually flown to Prestwick free by SAS, and on to Shetland by BEA, and came to Fair Isle on the mailboat, *Good Shepherd*, on 9 July.

The birds are being looked after by Roy Dennis and Tony Mainwood, of the Bird Observatory, who will continue studies of behaviour and feeding habits after their release in mid-September when the birds will be fully fledged and, we hope, able to fend for themselves.

The first stage of this exciting experiment has been successfully concluded; the next will begin in September when, one by one, the birds are released. We can but hope that the abundant food supply will persuade them to remain on the island; but it will take four years for them to reach breeding maturity, and a lot can happen in that time.

Apparently a Norwegian landowner, T. Ulsund, provided one of the eaglets, asking that it be presented to our Queen. That same year, Andrew Macmillan celebrated the event in his editorial for *Scottish Birds*:

> Of a dozen species which have ceased to breed in Britain since 1800, five have been regained this century, three more breed sporadically, and only four (Great Auk, Great Bustard, White-tailed Eagle and Kentish Plover) are still absent. Probably only the White-tailed Eagle of these four could possibly be re-established as a regular breeder...It is therefore interesting to learn of another attempt to reintroduce this species...It will be an exciting moment if one day we can again watch this huge, heavy eagle soaring along the remote cliffs of the north and west.

Roy Dennis then took up the story in two articles for the *Fair Isle Bird Observatory Report* in 1968 and 1969. Two temporary cages had been erected on a hillside near sea cliffs known traditionally as Erne's Brae. It was here, as the name suggests, that Sea Eagles once bred on Fair Isle. The two cages each consisted of two compartments four metres square, and were provided with a small box-shelter and a perch. The birds were thus kept as pairs, separate, but within sight of a sibling.

One female, named Torvaldine, precocious from the outset, fed boldly from the hand; the other, Ingrid, preferred to take food placed on the ground in front of her. The male, Jesper, was shy and cowered in one corner of his cage but after his first night was found to have fed himself. All three were about seven or eight weeks old and had reached maximum weight, although had yet to complete feather development. The birds soon began to exercise and the first flight inside the cage was made on 20 July. The latecomer, Johan, was younger than the rest but soon caught up. About 350 g of meat or fish was provided to each bird every day but not all of this was eaten. Later the eagles became more shy and restless and, so that they would not become tame, visits with food were made only every two or three days.

Prior to its release, each bird was caught for weighing and measuring, each being given a metal numbered ring on one leg, with an individual colour ring on the other. Ingrid was the first to be set free, and on 16 September the wire wall was pulled back, using a long cord, thus allowing the bird to leave undisturbed and in her own time. With a seeming flair for anti-climax, Ingrid decided to walk to freedom and would spend the next 30 minutes scaling a small hillock nearby. It was only when a raven landed briefly beside her that she took to the air for her maiden flight. Her next flight was more prolonged, although she had to endure a mobbing from a passing peregrine.

Within a couple of days Ingrid had become more adventurous but still suffered persistent attacks from other birds, mainly crows and gulls. Although food had been dumped at a conspicuous site for her use, she preferred to search the beaches for carrion. On 21 September, after five days of freedom, she was observed eating a freshly dead oystercatcher but it is not certain whether she had managed to kill it herself.

Tony Mainwood feeding a young Sea Eagle on Fair Isle, 1968 (Norman Elkins)

Roy remembers how he was told by Geordie Stout, one of the Fair Isle crofters with an impressive ornithological pedigree, 'I have just seen one of your eagles, Roy. It looks like a barn door flying around'.

And so the description entered the lexicon of ornithological terms. On 2 October Ingrid was joined by Jesper, and then by Johan two days later. All three Sea Eagles regularly returned to the empty cage to feed on rabbits and dead birds left there as a food dump. When Torvaldine was released on 20 October, a new dump had to be located on the cliff top. During this week, in a spell of bad weather, Johan disappeared, having probably wandered out to sea; he was not seen again.

The three remaining Sea Eagles were often seen displaying and calling together during the ensuing winter. The male would swoop down on a female, who would roll over with spread wings to present talons. Occasionally a pair would grapple and tumble out of the sky a short distance before disengaging. They continued to utilise the dumps but were also beginning to find food for themselves. An eagle pellet containing fish bones was found, while the birds were seen feeding on dead birds or an occasional seal carcass on the shore. Torvaldine and Jesper continued to utilise the food dumps while, from 6 March, Ingrid had begun to lead a more solitary existence. Apparently in good condition, she was last seen on 12 April 1969, seven months after her release. She probably left the island during a spell of clear weather when Shetland to the north and Orkney to the south were both visible on the horizon.

The first cast feathers were found on 9 April after which the wing and tail moult became obvious. At this time the eagles sought refuge on the cliffs and spent less time in the air. The lambing season was by then in full swing and although once or twice the eagles hovered curiously over a ewe with her lamb, they made no attempt to attack. Nor were they seen to chase rabbits, and these displayed no fear of eagles overhead. Torvaldine and Jesper must have been, by then, largely self-sufficient and carrion was obviously an important constituent of the diet. The first indication of an actual kill was on 8 May when one of the eagles was seen to catch a fulmar in flight. It released it almost at once, but later caught another which was held for longer before it was let go. On 20 May, Torvaldine was seen carrying a glass bottle in her talons, which she had presumably snatched from the surface of the sea in mistake for a fish. Thereafter the remains of freshly killed fulmars began to be found on the cliff tops.

In clear weather the eagles often soared to 700 metres or more and even ventured 4–5 km out to sea. Eventually, during the second week of June, about eight months after release, Torvaldine left the island. Only Jesper now remained and, despite his heavy wing moult, he continued to catch fulmars, as well as two young shags taken from a nest. He remained hidden amongst the cliffs for the next few weeks although the remains of a young fulmar, a shag and a gull were found on the shore. On 19 August he was flushed out of a sea cave but, unable to fly properly, fell into the water. He drifted ashore and was caught. Jesper was fat and well-fed but his plumage was smeared in fulmar oil. It is thought that he had tackled young fulmars on their nest ledges, only to be spat at repeatedly before finally despatching his victim. He was set free again but probably died soon afterwards, being last seen on 28 August.

Jesper had survived ten months in the wild and died in rather unexpected circumstances. Although several other predators have since been known to have become contaminated with fulmar oil, this must have been an unusual occurrence. Sea Eagles can, and do, regularly catch fulmars, either from the surface of the sea or in flight, apparently without any risk of being spat at. It proved Jesper's undoing that he should opt for an easier alternative by walking up to fulmar chicks which looked deceptively helpless on their nest ledges.

Young fulmar spitting oil

No other eagles were released on Fair Isle, yet the project cannot be dismissed as unsuccessful. It is possible that one or two survived, having already demonstrated that they were capable of looking after themselves. From 1968 there were several vague and unconfirmed reports of Sea Eagles in Scotland, some later ones involving adult birds and therefore possibly of Fair Isle origin. As the release programme involved so few birds and lasted for only one year, a succcessful outcome was unlikely, yet it proved the feasibility of a method that could be employed in any further reintroduction attempts.

I was to be involved in the next reintroduction in Rum from 1975 to 1985. Sadly, after an unsuccessful kidney transplant, George Waterston died in 1980. Pat and Morton are also now gone. They would all, however, be delighted at the success that has been achieved. I have returned to Fair Isle quite a few times but am still waiting to see Sea Eagles breed there. George would be disappointed. Despite their suitability as Sea Eagle habitat, both Shetland and Orkney remain vacant, but it may just be a matter of time.

Chapter 2
Sea Eagle facts

Methinks I see her as an eagle
mewing her mighty youth
and kindling her undazzled eyes
at the full mid-day beam.
John Milton (1608–1674)

The everyday term 'eagle' is not a scientific one, merely an arbitrary name applied to any large bird of prey. Thus eagles do not form a distinct taxonomic group but rather embrace a spectrum of sixty or so species, not all closely related.

Nowadays seven species of New World Vultures and Condors are considered closer taxonomically to the storks and herons, sharing common features in skeleton, behaviour, even the habit of cooling themselves by defecating on their legs! That they look so like Old World Vultures is not because they are related, but because they share adaptations for a similar lifestyle – a classic case of convergent evolution.

The Old World Vultures are classified in the order Falconiformes, which contains three sub-orders – the Falcones (about 60 species), the Sagittarii (a single species, the Secretary Bird *Sagittarius serpentarius* from Africa) and the Accipitres, the largest, with some 250 species. Of its two families, the Pandionidae contains but a single species – the Osprey (*Pandion haliæetus*). The latter, with its specialised fish-eating habits, shares some adaptations with the sea eagles – another example of convergent evolution.

Within the third family of Accipiters, the Accipitridae, two lines seem to have emanated from primitive kites. One includes snake eagles, harriers, hawks, buzzards and the true or 'booted' eagles (so called because their legs are feathered down to the toes. This group contains the Golden Eagle *Aquila chrysaetos*).

The habits of some kites – and the Brahminy Kite (*Haliastur indus*) in particular – indicate a second line within the Accipitres which leads to the sea/fish eagles. This kite feeds near water where it will take crabs, small snakes and fish, as well as some carrion. Its courtship and plumage are reminiscent of sea eagles.

Within this second group are also classified the Old World Vultures, their affinities being suggested by the curious Vulturine Fish Eagle (*Gypohierax angolensis*). This bird is sometimes

Golden Eagle

Brahminy kite in captivity

called the Palm-nut Vulture, reflecting the taxonomists' dilemma in deciding to which sub-family the bird belongs: if only for convenience it is most often classified with the sea eagles. Presumably as an adaptation to its feeding upon messy oil palm fruits, the bird's bright orange face is devoid of feathers and so it bears a superficial resemblance to the Egyptian Vulture (*Neophron percnopterus*).

Palm-nut vulture in captivity

Egyptian vulture in captivity

Table 1: Classification of the Falconiformes

ORDER: FALCONIFORMES

4 sub-orders:

CATHARTAE (New World Vultures and Condors – 7 spp., now considered to be more appropriately classified with storks and their kin, the Ciconiformes)
FALCONES (Falcons and Caracaras - 61 spp.)
SAGITTARII (Secretary Bird - 1 sp.)
ACCIPITRES (c. 250 spp.):

2 families:

Pandionidae (Osprey - 1 sp.)
Accipitridae

8 sub-families:

Milvinae (Kites - 32 spp.)

Aegypinae (Old World Vultures - 15 spp.)
Circaetinae (Snake Eagles - 15 spp.)
Haliæetinae (11 spp.) Genus: *Gypohierax* (1 sp.) *Ichthyophaga* (2 spp.) *Haliæetus* (8 spp.)
Circinae (Harriers etc. - 13 spp.)
Accipitrinae (Hawks etc. - 58 spp.)
Buteoninae (Buzzards etc. - 53 spp.)
Aquilinae (Eagles - 37 spp.)

The two species of fishing eagles *Ichthyophaga* from southeast Asia are also classified with the sea eagles. Where their ranges overlap, the Lesser Fishing Eagle (I. *nana*) frequents faster streams in forested areas and often at greater altitude; the Grey-headed Fishing Eagle (I. *ichthyaetus*) has a conspicuous white base to its tail and is the more vocal of the two species. Both nest near the tops of trees and lay between two and four eggs.

Until recently, it was accepted that the bones of a Sea Eagle had been found on the Chatham Islands, east of New Zealand. Henry Ogg Forbes had collected many fossils there early in 1892, during a Canterbury Museum expedition. A few years later Forbes took all his fossils to England where in 1952 they were re-examined and some bones – three tarso-metatarsi, two pelves and a scapula – were identified and attributed to the genus *Haliæetus*, related to White-tailed and Bald Eagles. This seemed strange to say the least, the bones belonging to the Northern Hemisphere species pair, and not to any of the more southerly, tropical *Haliæetus* species. Another examination some years later even considered them a new species belonging to the genus *Ichthyophaga* – 'because of the position of the outer proximal foramen' in the leg bones – but this was soon dismissed and the species was transferred back to *Haliæetus* to be accepted as such, albeit uncomfortably, in respected ornithological texts.

Despite many fossil finds since then, no other eagle bones have ever been uncovered at that Chatham site. Finally, in 2002, New Zealand researchers Trevor Worthy and Richard Holdaway resolved the issue. The bones were indistinguishable from the Alaskan race of the Bald Eagle – because that is exactly what they were, having been found in Canada, and not on the Chatham Islands at all! Apparently, later in the 20th century, the bones had been re-labelled, along with much material that Forbes had later collected from Native American middens in British Columbia, and assumed to belong to his Chatham assemblage. The Chatham Islands sea eagle *Haliæetus australis*, then, has never actually existed.

Another mystery surrounded a brief description by F. Wurmberg in 1787 of a bird from the coast of Java, which entered subsequent ornithological texts as the Maritime Eagle *Haliæetus maritimus*. It is likely that this was in fact a White-bellied Sea Eagle and is now laid to rest as a 'doubtful and invalid' taxon.

Nonetheless, *Haliæetus* is possibly one of the oldest known genera of living birds. A left tarso-metatarsus has been recovered from early Oligocene deposits of Fayyum in Egypt and dated at about 33 million years old. It is similar in general pattern and some details to that of a modern sea eagle. The genus was certainly well established by the middle Miocene, some 12 to 16 million years ago. A fossil *Haliæetus piscator* has been described from Upper Miocene deposits in France. All the remaining eight species of this most ancient genus *Haliæetus* are still extant. For some unknown reason one characteristic of the genus is that they all have a fused tarso-metatarsal joint in their middle toe. They are, listed in increasing order of size:

Sanford's Sea Eagle *H. sanfordi*
White-bellied Sea Eagle *H. leucogaster*
Madagascar Fish Eagle *H. vociferoides*
African Fish Eagle *H. vocifer*
Pallas's Sea Eagle *H. leucoryphus*
American Bald Eagle *H. leucocephalus*
White-tailed Sea Eagle *H. albicilla*
Steller's Sea Eagle *H. pelagicus*

Collectively, the eight sea eagle species are aquatic or coastal in habits, and nest in trees, on cliff ledges or sometimes even on the ground. Usually two or three large white eggs are laid. Fish features to a greater or lesser extent in the diet but the sea eagles may reduce possible competition with the Osprey by being less specialised and more ready to take waterbirds, carrion and sometimes mammal prey. They may also pirate food from other species, not least from the Osprey itself.

Again, typically, all *Haliæetus* species except Sanford's Sea Eagle display prominent patches of white in the plumage, notably on the tail but also on the head, neck or underparts. In adults the beak and eye are usually yellow but remain dark in the more ancient lineages. It seems that sea eagles and fishing eagles originated in the general area of the Bay of Bengal at a time, some 30 million years ago, when India was about to 'collide' into Asia. A group of four tropical, Southern Hemisphere species arose soon after, when the African and Madagascar Fish Eagles diverged. As the vast expanse of shallow ocean between India and Asia closed, the Pallas's Sea Eagle acquired its essentially land-locked distribution, a perculiar feature among sea eagles. As a result it is less dependent upon water and occurs at altitudes of up to 5,000 m in some parts of its range. Pallas's Sea Eagle is the enigmatic member of the genus, seeming to be closer to, and perhaps an early offshoot of, the northern species lineage which evolved at a later date. It did not

Pallas's Sea Eagle in captivity

acquire the hefty yellow bill and yellow eye of this group, retaining instead the smaller black beak and dark eye of the tropical sea eagles.

> Wednesday, 27 March 1996, Lake Baringo, Kenya. . . .The main highlight for me were the African Fish Eagles, and we saw one or two perched near a nest, but they were too full to respond to the boatman's whistles and offer of fish. Raphael had inserted pieces of balsa wood inside the fish to make them float, so we could retrieve the fish to try again. At last, at another point along the shore, a fish eagle came down but, not quite knowing what to expect, I missed the chance to take photographs. The eagle took our fish back to a tree to consume, so we went off in search of another. This time the bird came directly towards the camera but I found it hard to focus manually. Two other Fish Eagles proved better. I had a machine wind and – in these pre-digital days – I had no idea what I had managed to capture until the film was developed a few weeks later. But I found that using a shorter tele and focussing on the fish rather than the bird, was probably going to be the best strategy. It was a spectacular thing to witness from such close quarters, and the skill of the eagles in snatching fish from the surface of the Lake was indeed impressive. (The balsa wood was discarded at the eagle's plucking post).

The Latin name of the African Fish Eagle is *Haliæetus vocifer* and it is indeed very vocal. It is often referred to as 'the voice of Africa' and no-one who has encountered the species in its native habitat would quibble with this. To catch the skirl of an excited fish eagle along a riverside, beside a lake or marsh makes a memory to take home, haunting whoever hears it long after they have left. This very vocal nature is, in fact, characteristic of the whole genus. In complete contrast, the Golden Eagle is rarely heard. As with any bird, the fledglings can be pretty noisy when begging for food but, apart from a hungry imprinted juvenile which I had in captivity for a short time, I have only once, faintly, heard wild golden eagles calling when I was watching them copulate near the eyrie.

African fish eagle, Okavango Delta 2005

Six members of the genus *Haliæetus* fall into three obvious pairs constituting, for some reason, one white-headed representative and the other tan-headed. The white-headed African Fish Eagle is widespread south of the Sahara contrasting with the critically-endangered Madagascar Fish Eagle, confined to the island's northwest coastal forest.

Similarly, the White-bellied Sea Eagle, with its dark beak and eye, only gave rise to the dark-headed Sandford's Sea Eagle comparatively recently, less than a million years ago. The first is widespread throughout southeast Asia as far as Australia, while the latter is confined to the Solomon Islands. Sanford's Sea Eagle seems to retain into adulthood an immature-type plumage of rufous and dark brown; its head is brown, its beak black and its eye dark brown. It is almost as though the species has assumed the vacant role of true eagles, inhabiting coastal lowland forest where it feeds on fish, birds such as pigeons and beach carrion, as well as mammals like the phalanger or cuscus, a marsupial possum, and fruit bats. The nest and eggs of this rare and localised sea eagle have yet to be described though it nests in trees just like other *Haliæetus* species.

White-bellied Sea Eagles in captivity

The tan-headed, Gray or White-tailed Sea Eagle forms the third species pair with the white-headed, black-bodied Bald Eagle. Their common ancestor probably diverged from other sea eagles some ten million years ago, perhaps even as early as 28 million years ago. More recently, in the North Pacific, they split into two lines, one population spreading westwards into Eurasia as the White-tailed Sea Eagle, and the other eastwards as the Bald Eagle into North America. The Bald Eagle is not, of course, bald: the name could be using the old word 'bald' meaning 'white', or else it might be a reference to the word 'piebald' reflecting white head and tail contrasting with the dark body. Its specific name *leucocephalus* is Latinised Ancient Greek for 'white head'.

Sea Eagle and Bald Eagle share the huge golden beak and yellow eyes of the third northern species, the massive Steller's Sea Eagle, with which they may overlap in parts of Siberia. But, like the Pallas's Sea Eagle, Steller's (named after the Russian naturalist Georg Wilhelm Steller [1709-1746]) does not form a natural pair with any other living species. A dark form has been described from Korea and was first thought to be a distinct species, Heude's Eagle *Haliæetus niger*; it then became a subspecies of Steller's Sea Eagle *Haliæetus pelagicus niger* until, in 2001, a female was hatched in captivity from regularly-coloured *pelagicus* parents but exhibiting this dark colouration.

Steller's Sea Eagle is usually said to be the third largest eagle in the world behind only the South American Harpy Eagle *Harpia harpyja* (males 4.0–4.8 kg, females 6.0–9.0 kg) and the Philippine Monkey-eating Eagle *Pithecophaga Jefferyi* (4.7–8 kg). I am not sure on whose authority this judgement was made, but it has been recycled ever since. Even the latest *Handbook of the Birds of the World* (1994) and *Raptors of the World* (2001) fail to present a convincing body of measurements. The latter quotes wing lengths and weights for Steller's Sea Eagle (max. 620 mm and 9 kg respectively) that are at least as much, if not greater, than either Philippine (max. 612 mm and 8 kg) or Harpy Eagles (max. 626 mm and 9 kg). Being forest eagles, both have broader wings and longer tails than do Steller's, of course, while the Philippine Eagle in particular has impressively thick legs. A captive bird I once saw in Berlin Zoo had

Bald Eagle, Vancouver Island, British Columbia

Steller's Sea Eagle, Edinburgh Zoo

quite the most massive tarsi that I have seen in any flying bird. Despite that, though, for me, Steller's Sea Eagle remains one of the most magnificent birds I have ever set eyes on, even if that has only been in captivity (in Berlin and Edinburgh Zoos). I remain sceptical of any ranking that claims it is only the third largest eagle in the world, while the White-tailed Sea Eagle, reputedly fifth, can still sit proudly alongside these top three in whichever sequence one decides to place them.

There once existed an even larger eagle, related to the 'true' eagles *Aquila*: Haast's Eagle *Harpagornis moorei*, known from numerous fossils found in South Island, New Zealand. Its wings were short and broad with a long tail, typical of forest eagles. The wing bones, however, are shorter than might be expected, so the species may have been evolving towards flightlessness. Some 30 per cent heavier than a Harpy Eagle, a large female Haast's may have had a wing span of 2.5 m or more and weighed as much as 14 kg. The males were smaller – 2.2 m and 11.5 kg – exhibiting a degree of sexual dimorphism that is typical of raptors that actively pursue their prey. It apparently preyed on the large flightless moas, some of whose pelvic bones have been found punctured by the eagle's powerful talons. It may also have taken carrion. Both eagle and moas are now extinct, although rock paintings of a giant eagle exist, dating from the 13th or 14th century, so it may have persisted into comparatively recent times (Worthy and Holdaway 2002).

There are no obvious differences in plumage amongst the sexes in adult Sea Eagles. Size, therefore, is the most useful distinguishing characteristic. In the animal kingdom it is often the male of a pair that is larger than the female. In most birds, if they show any difference at all, it is the male which is the larger. Birds of prey are usually the opposite. The wings and tail of female Sea Eagles are some 6–7 per cent longer than those of males, the beak some 8–10 per cent longer and the tarsi about 14 per cent thicker. Females, furthermore, are about 20 per cent heavier than males. Weight, on the other hand, is a highly variable statistic when one considers that a kilogram or more of food can be consumed at a single meal; weight can also

Male (left) and female Sea Eagles, a few months old, Rum

vary both daily and seasonally. I had a female Sea Eagle in captivity for a time which reached a weight of 7.5 kg.

Dimorphism in size is not pronounced in the Sea Eagle. It feeds on less active prey than most other birds of prey and, indeed, true carrion-feeders exhibit almost no size difference at all. On the other hand, a species such as the Sparrowhawk *Accipiter nisus*, specialising in fast-moving avian prey, has females that are twice as large as males. Such pronounced size dimorphism is shared by owls and even skuas, with similar predatory habits. Ian Newton (1979) discussed several ecological and behavioural explanations which have been postulated to account for reversed size dimorphism, but no single explanation may provide the complete answer.

The biology of the various members of the genus *Haliæetus* is remarkably similar. All are associated with water, although to varying degrees. Pallas's Sea Eagle can venture far inland, often being associated with lakes and rivers, but might even be found on arid steppes. The African Fish Eagle occurs far inland throughout sub-Saharan Africa, up to 4,000 m above sea level, but nearly always near freshwater lakes, rivers and marshes. Its sister species, the Madagascar Sea Eagle, is more confined to the coast while all the other species are more catholic in their choice of habitat, quite prepared, at times, to frequent inland lakes, rivers and marshes. Indeed, Steller's Sea Eagle favours heavily wooded river valleys. Like most *Haliæetus*, Steller's Sea Eagle may build large stick nests in either tall trees or cliff ledges. Some might even nest on the ground on offshore islands where they are free from disturbance. Only the African Fish Eagle tends to shun cliff sites, while its Madagascar cousin will nest on either crags or trees on the coast.

Throughout much of its range the White-tailed Sea Eagle also tends to be associated with the coast, especially in Norway, Iceland and Greenland, where it often nests on cliffs or low rock faces, even on the ground on undisturbed offshore islets. As a result up to 60 per cent of eyries may be considered easily accessible to humans (Christensen 1979). One enterprising Norwegian Sea Eagle pair chose to build on top of a 7 m-high seamark on a coastal shipping route (Willgohs 1961). In many countries further south in Europe, though, it may venture inland to breed, but invariably close to lakes, rivers or marshes. In East Germany, for instance – and probably fairly typically for most of central Europe – 75 per cent of Sea Eagle eyries were situated within 3 km of a lake (Oehme 1961). Eighty per cent of these eyries were

White-tailed Sea Eagle eyrie in conifers, Sweden (Björn Helander)

in forests but the eagles exhibited a distinct preference for nesting on wooded islands or promontories, and nests were usually towards the edge of an open space – a clearing, marshy ground or even agricultural land. Nest situation will vary from place to place, of course, but Gunther Oehme knew of only one nest on a rock ledge, on the chalk cliffs of the Baltic island of Rügen. Of the 177 nests he examined, 65 per cent were in mature pines, 22 per cent in copper beech, 85 per cent in oaks and the remaining three nests in alders. The eagles tended to prefer the tallest trees with the eyries placed 8– 30 m from the ground.

In contrast, Willgohs (1961) visited 98 nests in Norway, only eight of which were in trees – two in pines, five in birch and one in willow – but his surveys tended to be conducted from a boat, visiting coastal islands along his route. My friend Harald Misund, working mostly on foot over the same ground around Bodø, finds many more eyries in trees which Willgohs would have missed. My own examination of the early naturalists' accounts from Scotland indicated only 20 per cent of Sea Eagle eyries here were in trees; the species was not always recorded but four were in pines, and one each in alder, rowan and birch. This may have been a consequence of the deforestation that has stripped the Highlands for centuries.

Less nest material is necessary to construct a suitable nest on rock ledges but through time, as more and more material is added, tree nests may achieve considerable proportions. In Ohio in 1925, after 36 years of use, a Bald Eagle nest crashed to the ground during a violent storm. It measured 2.6 m across, had been 3.7 metres tall, and was estimated to weigh about two metric tons, approximately as much as a small car. The world's largest nest was found in Florida, measuring 2.9 m across and 6.1 m tall. I wonder if anyone has ever been able to measure the nest of the largest eagles in the world, the Harpy or the Philippine? Willgohs photographed a remarkable White-tailed Sea Eagle eyrie placed in the cleft of a tree where, over ten years, it reached a height of 3 m. Another reached twice the height of a man until the tree could support it no longer and broke. The weight of winter snows often prove the final straw. One eyrie comprised 600 kg of sticks and other nest material, while John Colquhoun (1888) referred to another in Scotland as containing 'not less than a cartload of sticks'. With eyries often 3 m or more in diameter, a Sea Eagle nest in Greece reputedly served as a refuge for a notorious local bandit!

It is not unknown for Sea Eagles to use the nest of other species such as Black Kite, Buzzard or Raven, while pairs have been seen to evict Ospreys and Red Kites. In Scotland, golden eagles took over many Sea Eagle nest sites but probably only after the latter had become extinct. On the other hand, and as we'll see later, there is now evidence of the reintroduced Sea Eagles moving back in, and perhaps even evicting their more aggressive cousin. There are American reports of Great Horned Owls nesting within an occupied Bald Eagle nest, while one pair of owls even succeeded in evicting the eagles altogether.

Bald eagle nest, Vancouver Island, British Columbia

An interesting situation is recorded from Mecklenburg in East Germany (Deppe 1972). After 30 years' continuous use, a White-tailed Sea Eagle nest had to be abandoned due to timber operations nearby. When the birds returned to it two years later, they found a pair of peregrine falcons in residence. Undeterred, the eagles built a new structure above it. Many aerial battles ensued but the falcons failed to expel the eagles who only gained respite from attack once they had landed on their own nest. This uneasy situation persisted throughout that year (1947) and the next, when both eagles and peregrines succeeded in rearing one youngster each. Johan Willgohs showed me a cliff face in northern Norway with a pair of Sea Eagles in residence, ravens nesting higher up and a pair of peregrines near the top.

The normal clutch is two large white eggs, sometimes one or three and rarely, in some species like the African and Pallas's Fish Eagles, four eggs. Referring to the White-tailed Sea Eagle in Scotland, the Rev. F.A.O. Morris (1851) was moved to comment how 'the eggs… by a merciful provision are few in number'. Sea eagle eggs are described as blunt-ovate to almost round in shape, rough in texture and white in colour, though they can quickly become stained during incubation. Single eggs occur not infrequently, sometimes due to the accidental loss of the other(s), but it is probably true that young inexperienced

White-tailed Sea Eagle nest in Norway (Johan Willgohs)

Sea Eagle clutch in Swedish eyrie (Björn Helander)

females might produce only one. Three egg clutches are not uncommon and apparently, early last century in Skye, the shepherd Duncan Macdonald once found four eggs in a Sea Eagle nest (Harvie-Brown and Macpherson 1904).

Other four-egg clutches are known from Poland, Romania and Norway. Willgohs and I were shown an eyrie by a local fisherman that had contained four eggs, although the huge nest collapsed out of the tree before incubation had been completed. It is usually obvious from brown staining if any infertile eggs had survived in the nest over winter, escaping crows which are normally quick to mop up any left unattended. Instead I wonder if these four-egg clutches are the product of two females laying in the same nest in the same season. Trios are not uncommon and several instances are known since our reintroduction, though usually one female emerges dominant. Recently in Scotland at least one four-egg clutch resulted from such a trio; the nesting attempt failed.

Incubation is undertaken by both species, but the female often takes the larger share, sitting for a period of 38–40 days per egg. They are laid several days apart so, as with most raptors, one chick is larger than its siblings. It is not uncommon, though, for twins to fledge. Steller's and African Fish Eagles often rear only one, while Madagascar Fish Eagles rarely fledge more than one, while also seemingly prolonging the fledging period by several weeks, from a norm of about ten weeks. Not only does the Madagascar Fish Eagle exhibit sibling rivalry to reduce its brood to only one eaglet, but this species is unique within the genus, indeed among eagles in general, in breeding in co-operative groups. Ruth Tingay (2005) found that up to 40 per cent of the nests had extra adult helpers, more often males than females, sometimes as many as five birds, but they may not necessarily have been related to the established breeding pair. Sometimes analysis of DNA reveals that an extra male has sired the progeny, rather than the established territory holder.

With only 100 pairs or so, the Madagascar Fish Eagle is the rarest of all eagles, and indeed one of the seven most endangered raptors in the world. As a result it has attracted special research and management by the US conservation organisation known as the Peregrine Fund.

> Madagascar. In February 2012 I was guiding on a small expedition cruise ship Clipper Odyssey in the Indian Ocean. Only three days previously I had seen an African Fish Eagle perched on a tree on Islas Rolas, a tiny island just north of Kilwa Kisiwani on the Tanzanian coast. Now in Madagascar, about 400 km to the east, we were cruising off the small island called Nosy Anjombavola in the Nosy Hara National Park. Ruth Tingay of the Peregrine Fund had alerted me to the chance of seeing Madagascar Fish Eagles here and sure enough, we found the first perched on a small crag, watching us. It took off round the corner in a flurry of Palm Swifts, but we found it again and then saw two or three others around a smaller island nearby. One of our guides claimed that there were eight pairs around here but Ruth thinks this unlikely. I did not get close enough for a good photograph but it had been one of my aims to see this rare eagle in the wild. Mission accomplished.

The eight *Haliæetus* species are also remarkably similar in their hunting techniques and catholic in their diet. Fish are usually the favoured prey, snatched from the surface by the talons after a shallow dive in flight or from a perch nearby. White-bellied Sea Eagles will also take sea snakes and small turtles this way and, like their sister species Sanford's, will disturb and snatch fruit bats from roosts. Water birds, especially diving birds and/or seabirds, are also common prey of all *Haliæetus*, with small mammals, reptiles and amphibians regularly caught live. Carrion also features in the diet of several species, though, and Pallas's have even been observed feeding on human corpses. White-tailed Sea Eagles have also been described doing the same, on Anglo-Saxon battlefields. Sanford's Sea Eagle is often said to have the feeding habits of a forest eagle yet it still relies heavily on fish. All species may indulge in discards from fishing boats or harass otters or piscivorous birds such as herons, ospreys and even other *Haliæetus* species. Steller's, Bald and White-tailed Sea Eagles in particular will gather during salmon runs or where fish are incapacitated in up-wellings, or around feeding pods of dolphins or orcas. I shall discuss the diet of the White-tailed Sea Eagle in more detail in a later chapter.

Female Sea Eagle incubating, Burg Gutenburg, Germany

The pair of fishing eagles in the genus *Ichthyophaga* feed almost exclusively on fish. The Lesser penetrates up forested river valleys, while the larger Grey-headed prefers coastal, lowland forest, rivers, marshes and estuaries. Both nest in trees but while the Lesser will lay up to three eggs, the Grey-headed usually lays only one or two. Incubation seems to be shorter, up to 30 days, but detail is distinctly lacking in this pair of little-studied, though not necessarily uncommon, species, just as it is with Sanford's Sea Eagle.

Steller's Sea Eagle eating a salmon, Edinburgh Zoo

Since the diet and breeding habits of the eight *Haliæetus* species are so similar it is not surprising that only one representative of the genus is to be found in Europe, North America, Africa and throughout southeast

Adult Bald Eagle, Vancouver Island, British Columbia

Asia and Australia. Rarely do their ranges overlap significantly. Curiously, however, the coasts and vast river systems of the South American continent are devoid of any sea/fish eagle, although there are fishing buzzards *Busarellus nigricollis*. Nor has the genus *Haliæetus* colonised Antarctica – for obvious reasons – New Zealand or the Pacific Islands. As we have already seen, fossils from New Zealand turned out to have been wrongly labelled.

A few bald eagles are recorded, albeit rarely, crossing the Bering Strait into Siberia while White-tailed Sea Eagles have been known in recent times to have made the reverse journey, retracing the ancient route by which the ancestral species colonised North America in the first place. Here, in isolation, it evolved into the Bald Eagle. The habits and morphology of the two species are so similar that they are considered to constitute what is termed a superspecies. Apparently, Alaskan bald eagles are larger than their southern counterparts. As yet, southern bald eagles – around the Gulf of Mexico, for instance – have shown no inclination to cross the isthmus of Panama to colonise South America, but instead move north to overwinter in the Upper States and Canada.

Curiously, bald eagles never colonised Greenland and it was from Europe that the ice-free areas of southwest Greenland came to be populated with Sea Eagles from the east. DNA seems to confirm that White-tailed Sea Eagles colonised Greenland from Iceland some 4,000–6,000 years ago. There have been no detailed biometric studies of the White-tailed Sea Eagle throughout its extensive range, but the largest Sea Eagles are known to occur in Greenland. One bird from there has recently been found to have a record wingspan of 8.3 feet, or nearly 3 metres. Indeed, one measured by a local minister in South Uist spanned 9 feet (3 metres) from wing tip to wing tip. In Eurasia, if egg dimensions are any indication, it is possible that there may be a clinal increase in size from south to north. This is certainly the case with bald eagles in North America, the Alaskan birds being larger and heavier than the Florida population (Table 1). However, some authorities still question whether the Bald Eagle and the White-tailed Sea Eagle do indeed merit division into full subspecies.

Table 2: Comparison of wing lengths and weights of Bald Eagles and White-tailed Sea Eagles (from Love 1983)

Race	Male	Female	Source
WING LENGTH (mm)			
H.l.leucocephalus	529 (?) 515–545	577 (?) 548–588	Friedman 1950
H.l.alascanus	589 (?) 570–612	640 (?) 605–685	Friedman 1950
	619 (8) 577–650	655 (8) 620–705	Dementiev and Gladkov 1951, Southern 1964
H.a.albicilla	595 (23) 575–625	653 (17) 635–690	Dementiev and Gladkov 1951
	641 (26) 590–695	685 (37) 650–740	Willgohs 1961
H.a.groenlandicus	647 (26) 624–665	691 (27) 660–727	Salomonsen 1950
WEIGHT (kg)			
H.l.alascanus	4.1 (4) 3.9–4.3	5.6 (7) 4.8–6.6	Dementiev and Gladkov 1951
H.a.albicilla	4.4 (12) 2.8–5.4	5.6 (22) 4.6–6.9	Willgohs 1961, Fischer 1979

The close affinities of the Bald Eagle and the White-tailed Sea Eagle are apparent not only in their almost identical habits but also in their appearance. My own experience of the Bald Eagle in Nova Scotia, British Columbia and California suggested it might be slightly shorter and narrower in the wing, with a longer but less pronounced wedge-shaped tail than the White-tailed Sea Eagle. Whilst the body plumage of the latter is normally described as brown, its head and neck can appear conspicuously pale, at times almost white. Some individuals can be surprisingly pale all over – 'a fine silvery white, without the slightest admixture of brown,' as Charles St. John (1849) described one eagle shot in Sutherland in 1848. He was told of another similar bird in its company, and it may have been one of these – 'pure white in colour' – which came to be preserved in the museum of Dunrobin Castle near Golspie. Another could be found in Lews Castle, Stornoway, in the Outer Hebrides, which Robert Gray (1871) described as:

> ...the finest British example of the Sea Eagle I have ever seen... This magnificent bird, which was killed in the island of Lewis, is distinguished for its great size and lightness of colour, being of a yellowish-grey all over.

There is an almost-white Sea Eagle preserved in a glass case in Inveraray Castle, which was claimed to have been one of the last in Britain. A gamekeeper killed it in a fox trap on Kilblaan Hill on 29 March 1915 and it is said to have weighed 16 pounds (7.3 kg). Whether this stuffed specimen has faded and bleached over time, or whether it really was as white as this, is probably impossible to say.

A stuffed, nearly-white Sea Eagle, trapped in 1915, in Inverary Castle (Dave Sexton)

Other similar individuals from Sutherland were recorded by John Wolley (1902) and Harvie-Brown and Buckley (1887). A silver female from Rum was shot on the Isle of Eigg in the Inner Hebrides in 1886 (Harvie-Brown and Macpherson 1904) while Harvie-Brown (1906) also recorded a pair of 'albinos' nesting in a tree in Loch Laidon, Perthshire, both parents and their young being 'alike of a pale dove colour or ash colour'.

This last record suggests there may be a genetic factor involved, although the common contention was that such pale individuals were of great age. John Wolley recalled one nesting on Loch Stack in Sutherland in 1848 which was 'so grey, it was supposed to be very old'. Certainly the last Sea Eagle in Shetland, according to the artist George Lodge (1946) was 'quite white and looks as white as a gull when flying'. Its primaries, though, showed pale brown. Another account in the RSPB's *Bird Notes and News* from 1918, which I have not examined, reported the death of this last Sea Eagle. It went on to claim that the bird 'outlived her mate for eight years, and had grown quite white'. On 23 April, probably in the same year, Lt. Commander I. G .Millais responded to this report in the *Daily Mail* by mentioning the bird as being 'a pale variety, or semi-albino.' Other reports seem to confirm the bird was apparently quite elderly, having been resident in the area for about 30 years.

The age of this very last Sea Eagle in Britain is not unusual. My friend Harald Misund ringed a Sea Eagle chick in northern Norway which was found dead 29 years and ten months later. Another that had been ringed in Finland in July 1980 was identified in Sweden recently, aged 30 years and seven months. In captivity, however, White-tailed Sea Eagles may live for well over 40 years. A captive Sea Eagle taken in Kirkcudbright in 1858 died in May 1900 at the age of 42 years, by which time the bird was totally blind (Maxwell 1907). There are even tales of captive eagles living to the age of 100, but they are no doubt all apocryphal. Seton Gordon (1955), for instance, repeated an unlikely story of a Golden Eagle reputedly shot in France in 1848; around its neck was a gold collar bearing the date 1750. However there is no guarantee that the bird that was shot was the same one that had been given the collar in the first place; such a valuable item would doubtless have been passed down to other eagles in that time. The oldest captive Golden Eagle which Gordon could legitimately claim had been 46 years old, while another is mentioned in North America that lived to be 60 years.

The plumage of any bird makes a considerable part of its bulk, weighing twice as much as the skeleton. Brodkorb (1955) diligently counted no fewer than 7,182 vaned feathers on a dead Bald Eagle, which collectively weighed 586 g. With an additional 91 g of down feathers, the plumage amounted to 17 per cent of the bird's weight. With wear and tear these feathers need to be renewed at intervals, but given the size of the bird and the energy required to accomplish this, it is not surprising that only about half of the plumage may be renewed annually, so moult patterns can often act as a useful, if short-term, means of distinguishing different individuals. Although I do not intend to describe this process here, information might be found in modern ornithological texts, such as *Handbook of Birds of Europe, the Middle East and North Africa*.

Sea eagles take about five years to moult into the full adult plumage, gradually acquiring the white tail, pale head, yellow eyes and beak. In its first year, the juvenile White-tailed Sea Eagle sports a coloration distinct from that of the adult. Its eyes are dark hazel-brown while its beak is black; both gradually become yellow over three years or so, as the bird

First year bald eagle (left) and Sea Eagle, British Columbia and Rum

matures. In its first year the bird is dark chocolate- brown, especially on the head and neck, but more rufous on the mantle, rump, scapulars and wing coverts. More white is apparent on the underside, giving the juvenile a conspicuously speckled appearance. The white bases to any of the feathers flash prominently when parted in a strong breeze, especially those at the back of the head. In flight the rufous-brown coverts on the leading edge of each wing contrast with the dark blackish-brown primaries and secondaries.

Having seen several juvenile bald eagles, both in captivity and in the wild, and being totally familiar with juvenile White-tailed Sea Eagles, I confess to finding it difficult to tell them apart. The Atlantic Ocean was always the most convenient distinguishing clue! At least it was until 18 November 1987 when a wet, bedraggled and painfully thin young eagle was found trying to feed at a pig trough in County Kerry. It could barely fly so was easily caught and, over the next two months, it was nursed back to health, gaining over a kilo to tip the scales at 4.8 kg. After much discussion across the Pond, and following careful measurements, identification finally boiled down to the degree of emargination on the sixth and seventh primaries! From this it was agreed that it was a six-month old Bald Eagle, the first ever recorded in Europe. (An adult reputedly sighted in Anglesey ten years previously had never been confirmed.) It was christened 'Iolar' – the Irish Gaelic for eagle – although, predictably, the press insisted on referring to it as 'Eddie the Eagle'. If it was indeed a Bald Eagle, had it flown across the Atlantic in a gale or was it, as some suspect, an illegal captive bird that had been released or escaped into the wild? (Pain 1988; Sexton 1988)

Amid a frenzied media circus, arrangements were made to fly the bird back 'home' by jet. Then, after 30 days in quarantine, Iolar was released in Massachusetts as part of a restocking programme involving bald eagles imported from Nova Scotia. Fitted with a radio transmitter its movements were followed by the project team for some weeks but I often wonder what happened to it after that. I would not have been surprised to read, several years later, of a White-tailed Sea Eagle resident in the United States.

Chapter 3
Sea Eagle fiction

There heard I naught but seething sea,
Ice-cold wave, awhile a song of swan.
There came to charm me gannets' pother
And whimbrel's trills for the laughter of men,
Kittiwakes singing instead of mead.
There storms beat upon the rocky cliffs;
There the tern with icy-feathers answered them;
Full oft the dewy-feathered eagle screamed around.

The Seafarer (7th-century Anglo-Saxon poem)

The eagle – 'king of birds' – has long been admired the world over for its majesty, strength and hunting prowess so it is inevitable that it should be adopted as a symbol of might and power. It features on many coats of arms and national flags around the world and has long held a special place in heraldry.

The legions of Rome marched under an eagle of silver, its wings outstretched and clutching a thunderbolt in its talons. Roman soldiers deemed it lucky to camp near an eagle's nest, although perhaps the breeding eagles would not have fully appreciated the military presence. Some legions even liked to bury an eagle's wing to protect the camp from storms. When a Caesar died, his body was cremated and an eagle set free amongst the flames. This was supposed to transport the emperor's divine soul aloft to the heavens. This whole cameo is reminiscent of the legendary phoenix, a supernatural bird of the Hittites in Asia Minor. When it grew old this mythical creature merely plunged into the flames, to emerge completely rejuvenated. This tale was perpetuated by mediaeval clerics who maintained that not only could an eagle look directly into the sun without harm, but that it could restore its youth by plunging into the sea. Many manuscripts and bestiaries depict such a myth and the eagle was, of course, taken as the emblem of St. John the Evangelist, who had looked upon 'the sun of glory' to acquire his faith and whose gospel Jeremy Mynott (2009) maintains to be the most 'soaring and revelatory'.

In the 12th century, Geraldus Cambrensis, the Welsh archdeacon and chronicler, wrote how the eagle of Snowdon was possessed of oracular powers and foreshadowed war. She would perch on 'the fatal stone', sharpening her beak before satiating her hunger on the bodies of the slain. Another 12th-century Welsh cleric, Geoffrey of Monmouth, who documented a history of the Kings of Britain, maintained that King Arthur is buried on Snowdon (one of many resting places claimed in the British Isles), his grave guarded by a pair of chained eagles. He also declared how sixty eagles would assemble annually at Loch Lomond in Scotland, to prophesy forthcoming events. The doyen of Scottish eagle watchers, Seton Gordon, related in his monumental book *The Golden Eagle* (1955) how the great eagles of Snowdon were said to breed gales and storms; indeed, an ancient name for this mystical mountain was Creig ian 'r Eryri, the rock of the eagles, probably golden eagles. Similarly Norse mythology attested how the wings of an almighty eagle generated the four winds in the world of men. It sat, with a falcon perched on its forehead, on the topmost bough of an ash tree which itself represented the world of the gods and was named Yggdrasil, the tree of knowledge. Beneath this tree coiled a serpent which was constantly at odds with the eagle. A nimble little squirrel was engaged to run up and down the tree trunk conveying insults between bird and snake!

13th-century manuscript in the Ashmolean Bestiary

Eagle and serpent symbolism is worldwide, signifying the struggle between the powers of the sky and the earth, of light and darkness. One legend from the New World tells of a tribe of Indians who trekked south in search of better lands. They eventually settled where they encountered an eagle perched on a cactus plant, holding a snake in its talons, a favourable omen. Thus the mighty Aztec civilisation was founded and this eagle, holding slippery prey on a prickly perch, is portrayed on the flag of the Mexican Republic to this day. Double-headed eagles are another common motif.

Besides an ancient role in heraldry, the eagle today has generated new symbolism. It is used as a trademark for many businesses, companies and products, for example. Indeed, playing on this association, Eagle Star Insurance agreed to support many conservation efforts, not least our reintroduction of the Sea Eagle to Britain. The Aros Visitor Centre outside

Eagle Star logo, now at the Aros Centre, Portree, Isle of Skye

Portree on the Isle of Skye houses an interpretive display about the island's Sea Eagles so it is fitting that the effigy at the entrance gate should be a cast-off from Eagle Star, appropriately repainted in Sea Eagle livery, of course!

Sea eagles as a group have long proved a popular icon. Sanford's Sea Eagle often features

Sea Eagle stamps

on the stamps of the Solomon Islands, as do other sea eagles in other countries. The African Fish Eagle is the national bird of both Zimbabwe and Zambia. Although there is a slightly larger 'booted' eagle present in Australia – the Wedge-tailed – several Antipodean rugby league teams invoke the White-bellied Sea Eagle with such names as Papakura Sea Eagles and Sunshine Coast Sea Eagles. The Manly-Warringah Sea Eagles incorporate a local place name, Warringah, being an Aboriginal word for an eagle's nest. One of our own rugby league teams, the Blackpool Sea Eagles, are obviously referring to the White-tailed, whilst the logo of the American football team, the Philadelphia Eagles, makes it clear that they invoke its sister species, the Bald Eagle.

The Bald Eagle is the national bird of the United States, and is depicted clutching in its talons thirteen arrows and a thirteen-leaf olive branch, reminiscent of the imagery of the ancient Roman Republic. I recounted a story in my first book that in 1784 Benjamin Franklin lamented that the Bald Eagle should be chosen, it being 'a bird of bad moral character; he does not get his living honestly; he is generally poor and very lousy.' This would seem unjust comment when one considers that the alternative being mooted was the Golden Turkey, 'a much more respectable bird'! In 1937 the ornithologist Arthur Cleveland Bent would implore that 'such a vain and pompous bird would have made a worse choice. . . Eagles have always been looked upon as emblems of power and valour, so our national bird. . .' – he added weakly – '. . . may still be admired by those who are not familiar with its habits.' It now turns out, against popular opinion, that this story is apocryphal. Franklin was merely using the two birds as a satirical device in a criticism of the fashionable Society of Cincinnati, whose ideals he saw as against those of the newly-independent Republic of the United States.

Symbols are expressions of desire, purpose and ideals, and the Bald Eagle motif has changed little since 1872, appearing on the national seal of the United States, on currency and coins, in countless governmental, commercial and private uses, pervading many aspects of American (and Canadian) life. It is said, too, that the Bald Eagle is the bird above all others that appears on the most postage stamps, not all of them American. The bird appears on the national insignia of the Philippines, for instance, and on some of their stamps.

Back with its sister species, our own White-tailed Sea Eagle, the world's first scheduled passenger service by flying boat was operated between Southampton and Guernsey in the Channel Isles by three Supermarine Sea Eagles. One of them crashed, another was rammed by a ship but the third maintained the service for five years until finally replaced in 1928. The Royal Navy's shore establishment near Londonderry in Northern Ireland used to be HMS *Sea Eagle* while, perhaps more ominously, the name Sea Eagle was given to a British anti-ship missile in the 1980s. The name is used on boats and seems

The Sea Eagle anti-ship missile by British Aerospace

to be coming in to more common usage as the reintroduced bird steadily impacts once again on our consciousness.

Eagles were sometimes trained to hunt and, as such, were regarded as symbols of great status. Traditionally anyone was allowed to keep and train Kestrels, the yeomanry kept Goshawks, the aristocracy Peregrines but only kings and emperors held the honour of flying eagles. Golden eagles were probably preferred, although it is said that King James I of Scotland (1406–1437) flew a White-tailed Sea Eagle when hunting Teal (Richmond 1959). Nowadays, eagles are rarely kept by private falconers because they are difficult and expensive to obtain, time-consuming to train and heavy to carry around on the fist. Commercial falconry centres, however, find eagles to be crowd-pullers and see them as profitable exhibits. It is not too difficult, then, to find a trained eagle to view close-up, always a thrilling encounter and for many people, not just children, an exciting educational experience. Personally, though, I have never been convinced by the conservation argument to justify condemning these birds to a life of captivity. At least they do get some exercise!

The practice of falconry probably originated in China some 4,000 years ago, possibly even earlier. Because training hawks and falcons was so time-consuming, it is likely that falconry was always more of a sport than a serious means of procuring food. Anglo-Saxon kings kept falcons, but it was not until the return of the Crusaders from the Middle East that falconry became popular.

Marco Polo found that the Kirghiz tribesmen of the Tien Shan Mountains in central Asia kept 'a great number of eagles, all trained to catch wolves, foxes, deer and wild goats'. It is difficult to believe but in 1923, one prize Golden Eagle was said to have captured fourteen wolves in a single day! It is doubtful that a Golden Eagle is sufficiently powerful to kill a wolf but it can at least hold one down until the owner gallops up to despatch the quarry with a knife. And the prey did not always come off the worse: one unlucky hunter lost no fewer than eleven eagles to a particularly ferocious wolf. Undeterred, the man patiently retrained a twelfth eagle and set out once more. This time the notorious wolf was slain. The triumphant eagle doubtless became more valuable than the normal going rate for a trained bird – one horse or

Falconry in 12th-century France

two camels. It is unlikely that the Kirghiz relied upon their eagles to provide regular sustenance, but rather flew them to show off their prize specimens. It took a lot of time and effort to fly an eagle at small prey such as hares or partridge for the pot.

> A gazelle on the other had provides meat and skin, while a good fox skin can be sold for the equivalent of £18 – more than a month's earnings in this part of the world. Falconry is not a sport to the Kirghiz. It is part of their lives and a means to an end – guns are strictly controlled and very rare.

Kirghiz tribesman with his Golden Eagle (Andy Miller-Mundy)

So wrote my friend, bird artist and falconer, the late Andy Miller Mundy, who travelled to Kirghizstan in 1995–1996 to see eagle hunters in action. In 1983, Andy, who lived at Harris in the Outer Hebrides, had shown me his Golden Eagle, a juvenile whose plumage had become matted with fulmar oil. His falconry training greatly facilitated the lengthy process of cleaning the feathers and rehabilitating the bird back to the wild. As it turns out it was Andy, too, who that year found the third Sea Eagle pair nesting in Scotland during the Rum reintroduction.

The late Andy Miller-Mundy with his rehabilitated Golden Eagle, Isle of Harris. (J A Love)

Just across the Minch, in Victorian times, Captain Macdonald of Skye once kept two Sea Eagles taken out of a nest on Dunvegan Head (Harvie-Brown and Macpherson 1904). They:

> ...became his familiar companions, descending from a great elevation to join him on his walks, answering his whistle, and retrieving the game [grouse] he had shot for their own larder... and often killed hares on their own account.

The naturalist William Macgillivray had been brought up in Harris early in the 19th century, not far from where Andy Miller Mundy later came to live. Macgillivray became Professor of Natural History at Aberdeen University, my old alma mater. His brother, Dr Donald Macgillivray, lived with his two sons at Eoligarry in Barra where they kept a Sea Eagle in captivity (Gray 1871). It would:

> ...follow them around and hover round their heads, perform the most graceful aerial

> evolutions, and scream with delight as if it thoroughly understood and enjoyed the expedition; and when an unfortunate rabbit showed itself the eagle would swoop down upon it with amazing rapidity and power.

Ultimately, the bird was mistakenly shot by a young visitor from Glasgow, 'to the great vexation of the entire household, with whom it had been a great favourite'.

Lord Lilford (1885) described an old female Sea Eagle that was taken from a nest in 1854 and was still alive and in good health in February 1890. Apparently it took charge of a brood of domestic chickens! He described the unlikely scenario of the eagle with two or three chickens on her back, while she was engaged in breaking up food for others running about at her feet. This is perhaps the same bird that Sir Herbert Maxwell (1907) had heard of being in captivity for 40 years while Macpherson (1892) knew of a young one from Borrowdale in the Lakes which was kept by the Bishop of Elphin; it acquired its white tail at six years old and died aged 19, but no date is given. Another trapped in December 1872 at Charterhall, Berwickshire (presumably a wandering immature) survived in captivity for nearly 14 years. It had an 8-foot wingspan and:

> . . . was exceedingly fond of trout and would scream with delight when fish were offered to it.

Sea Eagles, then, would appear to have made entertaining, if not always affable, pets. Robert Gray (1871) recounted an amusing tale of a Sea Eagle named Roneval, after a hill in South Uist where it was hatched:

> My friend [Colin] McVean has had a tame one for some years, which is not kept in confinement, but sometimes startles strangers by swooping past the windows. [McVean] says: 'He is a male and a very fine bird. I have had him now for four years, and he has assumed his white tail. He is allowed to fly about at large, but is not fond of going far, and will always come at the call of the kitchen-maid who feeds, him, and for whom he shows the greatest affection, and who can manage him even when in most ungovernable tempers. He has a particular aversion to small boys, and will fly at one going near him. The only animal he is afraid of is the pig, and to hear a pig grunt is enough to make him fly off, even if it should not be in sight. A well-dressed friend one day ventured to touch him with the point of his fashionable light umbrella, which so offended Roneval's majesty that he flew at the offending instrument and literally smashed it, breaking the stick and tearing the silk to tatters – the owner gladly escaping in unscathed broadcloth himself. . . Usually however he is affable enough, and does not more mischief than occasionally killing a hen or two if his own dinner is not served up punctually enough . . . This proves how very domestic this monarch of the cliffs may become, that though a short-winged flight would carry him to the illimitable freedom of the neighbouring sea cliffs and mountain tops, he has never been known to 'stop out of nights' more than once or twice during several years' residence.

In late July 1848, the naturalist John Wolley (1902) found himself in Caithness, too late to collect any eggs but managing to acquire a pair of live Sea Eagles, presumably chicks, from a nest on Dunnet Head. Wolley presented them to a relative, Mr C. Clarke of Matlock in Derbyshire:

> ...where I hope they may breed; but though five years old this season [1853] they have not yet quite completed the adult plumage.

His friend and biographer, Alfred Newton (1860), went on to describe how:

> ...a mass of rocks, perhaps in by-gone years tenanted by the other native species, was wired over, and the plan of the cage thus formed, having been brought to the knowledge of the late Secretary of the Zoological Society, suggested the first idea of the fine Eagle Aviary which now adorns the Gardens in the Regent's Park.

He later recounted that in this roomy cage, the birds lived contentedly for some five or six years:

> ...until one day it was found that the female had killed and eaten half her mate [presumably its sibling if they had come from the same nest]. On this she was transferred to other hands and when I last heard of her, was undergoing solitary confinement at Chatsworth – certainly an agreeable place of detention for a murderess.

John Wolley (1825–1859)

It is fascinating that so many of these early accounts refer to Sea Eagles, suggesting how much more familiar they were to local people than golden eagles. This was just as evident on the Isle of Rum. While living there I became fascinated by its human history which I researched into a book called *Rum: a landscape without figures* (2001). A particularly fruitful line of study was provided by papers from Lord Salisbury's ownership of the island. Amongst these I was delighted to uncover fascinating references to Sea Eagles. In July 1846, for instance, Lord Salisbury's manager on Rum, Alexander Mackenzie, had:

> ...two eaglets alive which Campbell [the shepherd] went down to the nest and took out. I will endeavour to keep [them] alive till his Lordship comes. They are of the largest kind.

By that last remark he means White-tailed Sea Eagles; he knew several pairs to nest on the island, and they were annually persecuted due to alleged depredations upon lambs. Two years later Mackenzie asked his employer if he could take a live eagle to send to an ex-officer friend in Yorkshire. In 1849 he obtained no fewer than three. 'They were all in one nest. I believe they have seldom more than two.' Again these are very obviously Sea Eagles. In October, Salisbury asked that the eagles be sent south to his estate at Hatfield:

> ...either two or three if you do not want any. They must be addressed to the Zoological Gardens, Regent's Park, London and Ward [the skipper of his boat] must take care to deliver them to the Guard of the Railroad, with a request that they may be sent to the station as soon as they arrive. You will take care to cut the feathers of a wing of each.

Mackenzie reported back immediately that the eagles had been despatched:

> The one I offered the gentleman I mentioned to your Lordship sent for it, but unfortunately it got loose and I had to shoot it to prevent it escaping. Captain Ward will be instructed to deliver them according to this order. The feathers are cut.

On 13 October, Ward reported to Salisbury that the eagles had been delivered. The following year (1850) an injured female Sea Eagle was sent to Lord Derby who had a small private aviary on his estate in Lancashire, now a safari park. His Lordship wrote to Lord Salisbury that the eagle's wounds 'will meet with every attention and not prove of any lasting inconvenience to it'.

This gives a little insight into contemporary Victorian 'natural history' and just how well known Sea Eagles were to people in the Highlands before the birds came to be persecuted to extinction in this country fifty years later. Rum would be one of the last places it bred.

Some old accounts indicate that 'eagles' were flown to catch live fish but they were in all likelihood Ospreys rather than Sea Eagles. However, the eyries of both species were sometimes robbed to procure fresh fish for the table. John Colquhoun of Luss (1888) knew of a shepherd lad in Blackmount who secured a good breakfast every day while the eaglets were rearing, simply by watching the feeding hour and robbing the eyrie. Another enterprising Highlander went so far as to put rings around the throats of eaglets so that he could be assured first of satisfying his own hunger. Osgood Mackenzie (1928), the owner and founder of Inverewe Gardens in Wester Ross, recounted how eaglets (in this case probably golden eagles) were tethered to delay their leaving the nest so that the adults would continue to bring them food – hares, fawns, lambs and grouse – which formed 'an agreeable variety at the shepherd's daily dinner of porridge and potatoes and milk'.

Sea eagles have long had an association with fishing, of course. In parts of Cambodia it is considered taboo to kill a fish eagle; the gods would punish anyone who did and also reduce the fish harvest. Shetland fishermen thought that as soon as a Sea Eagle appeared near their boat, fish would rise to the surface belly-up (Pennant 1774). This led them to hope that their catches would improve if they smeared Sea Eagle fat on their gear. I have seen fishing boats in Norway called *Havørna*, perhaps hoping that some of the bird's catching success will rub off or simply in sheer admiration of the bird's prowess.

Some North American Indians considered it unlucky to harm an eagle. Pacific tribes such as the Kwakiutl in British Columbia had a great affinity with bald eagles. They figured eagles on their magnificent totem poles and regarded eagle down as a sign of peace and friendship, sprinkling it before guests during welcome dances and other ceremonial occasions. Eagle feathers were used in many other such rituals and adorned masks, peace pipes and war bonnets. No warrior could wear an eagle feather, though, unless he had earned it by some act of courage in war.

Continuing the magical theme, the Anglo-Saxons believed that Sea Eagle bone marrow possessed miraculous curative properties (Armstrong 1958). It may well do – if anyone could ever acquire such a commodity, for bird bones do not contain marrow.

The Tartars feared that the talons could inflict incurable wounds, a not unreasonable belief, perhaps, given the potential of infection from a carrion-eater such as the Erne. Having been scratched and clawed by many Sea Eagles in my time, however, I can attest that it is not true – but then the Tartars did not have the benefit of antibiotics or tetanus injections. Curiously, the Faroese claimed that yellow eagle claws were a cure for jaundice (Bijleveld 1974) while some Malagasy tribes held that Madagascar Fish Eagle claws enhanced the potency of their magic concoctions. Ruth Tingay told me how some eagles are found missing claws. One of her study birds, to which she referred as Cut Off, lived for several years with only one leg,

Fishing boat Havørna in Bodø harbour, with Harald Misund (right)

and even bred quite successfully. Ruth admits she does not know how the bird came to lose its foot: perhaps the local fishermen had cut it off to retrieve an aluminium leg ring, mistakenly believing it was made of silver. Alternatively the bird could have been attacked by a Nile crocodile, or else lost its leg when it became entangled in fishing net.

There was also an ancient belief in Shetland (Ritchie 1920) that eagles raiding a farmyard could be prevailed upon by a charm to drop the victim: the witness was supposed to cast some knots in a length of string and utter a simple spell. It was prudently added that the successful outcome might only ensue some distance away, out of sight, no doubt saving the charmer some embarrassment. No less eminent a personage than the local minister from the town of Scalloway was said to have witnessed one such successful spell – in the absence of string, the aspiring charmer was seen to make do with his garter!

Eagles have long been vilified as predators of livestock. On the Lofoten Islands in northern Norway, bounties were offered for dead Sea Eagles. These were maintained until the 1950s and special stone huts were sometimes used to catch the birds. I was able to visit the island of Vaerøy where my friend Harald Misund showed me such an 'ørnehus'. It was in remarkable state of preservation, having been restored in the 1950s by the local teacher who employed it, not for the bounty, but to secure eagles for ringing and release. I

'Cut Off', the one-footed Madagascar fish eagle (Ruth Tingay)

The author entering an ørnehus *above the village in Vaerøy, Lofoten, Norway*

can do no better than quote at some length an early account given to David Bannerman for his monumental *The Birds of the British Isles* (1956):

> Hard by some eagle-haunted fell, the trapper dug a pit about a yard deep, around which he built a rough stone wall but of equal height, finally adding a roof of stone slabs. The entrance he closed with a large piece of turf. A little distance away was displayed some carrion, usually a sheep's entrails, secured by a cord which led into the hut. On a favourable morning the trapper would begin his patient vigil before daylight. Sometimes he had not long to wait before an eagle appeared. Even when most sharp-set, the bird would always break off its gorging at intervals to look around for possible dangers. Whenever it was so employed, the trapper carefully drew the carrion a little nearer to the hut. So far from showing any uneasiness as the lure moved, the eagle would tear at it more fiercely. If all went well the bird would follow its meal right up to the hut when the trapper, waiting until his quarry's attention was once more distracted, would seize it by the legs and drag it in. Finding itself in the utter darkness behind the turf curtain, the eagle would offer no resistance. If two or more eagles were at the bait together, it sometimes happened that when the first was captured the others continued feeding. With luck a trapper could catch two or three eagles in a few hours. But his patience often went unrewarded, and there he remained crouched in the hut until darkness fell.

On Vaerøy one man caught as many as 16 eagles in a single day. In such favourable localities, between 50 and 100 eagles might be caught in any one season. Towards the end of the 19th century an eagle would be worth three kroner – two as head-money and the third by selling the wings to be used as brooms.

In both Norway and Scotland, before firearms came into cheap and common usage, an automatic trap was employed to catch eagles. It was simply a narrow trench with walls partially built up with stones. Bait was placed inside and when the unsuspecting eagle jumped in to feed, it was unable to open its wings to escape. There is a valuable series of volumes describing all the parishes in Scotland, called the *Old Statistical Account*, which mentions dry stone walls with an opening at the foot of one wall for a noose. As soon as the eagle had gorged on the bait and tried to exit through the hole, it became ensnared in the noose, which eventually strangled it.

In remoter parts of the Hebrides, William Macgillivray (1886) described how a pit would be dug and deepened by a low wall of turf built around it, and roofed with sticks and heather. Poking his gun through a small aperture, the hunter would wait for an eagle to come to the bait, usually a sheep or horse carcass. One man shot five eagles in this way in a single winter. Another, an old shepherd on Skye called Duncan Macdonald, had a similar hide on a hillside in Glenbrittle which he baited with a dead sheep. He would enter in darkness and any eagle descending to feed at daybreak was 'shot in the head with his double-barrelled muzzle loader and BB shot'. On one occasion he shot three in one morning (Harvie-Brown and Macpherson 1904). William Macgillivray also described how broods were destroyed by lowering burning heather on to the nest from above. In his *A History of Fowling*, the Rev. H.A. Macpherson (1897) told of a 104-year-old Skye shepherd, Angus Macleod, who could recall killing 'more Eagles than any man of his acquaintance.' He would shoot the parents first and then descend into the eyrie with tar-soaked rags to burn or smoke out the eaglets.

A man catching a young Sea Eagle from an ørnehus on Vaerøy, Lofoten

From 1625 in Orkney (Low 1813) there was an Act:

> ... anent the slaying of the Erne. . . Whatsoever person or persones shall slay the earne or eagle, shall have of the bailzie [officer of justice] of the parochine [parish] qr it shall happen him to slay the earne or eagle, 8d for every rick [corn stack] within the parochine, except of the cottars who has not sheep; and 20 s to ilk persone for ilk earne's nest it shall happen him to harrie.

Similarly (Venables and Venables 1955) in Shetland:

> . . . are many ravenous Fowls, as Eagles, Ravens and Crows. In old times they so multiplied that the Fourde or Sheriff made an Act that whosoever at ye Head-Court brought in an Eagle's head, from each having sheep in that pasture he should have a merk . . '

By the 18th century the *Old Statistical Account* for Orkney and Shetland tells us how the reward had risen to five shillings a head. This was ultimately reduced to three and sixpence before the system ceased altogether in 1835. If you were lucky, claiming the bounty might not take a lot of effort. One Shetlander stumbled upon a Sea Eagle with its feet firmly trapped in the body of an enormous halibut. Similarly, a Highlander laid hold of one of a pair of Sea Eagles that tumbled into Loch Lomond whilst talon-grappling. One Faroese man even captured an eagle crouching innocently behind a rock as it sheltered from a snow storm – all he had to do was wring its neck.

Even allowing for the Old Scots conversion rate, the bounties being offered would appear to have been exceedingly generous; perhaps eagles were scarce and few people put in the effort to claim them.

Not infrequently in the past, human babies were claimed to have been taken by eagles, stories that might instantly arouse horror and suspicion, even fear. The accounts, though, cry out for critical assessment and few if any prove credible. Not surprisingly, I deliberately include them in this chapter dealing mainly with eagles and fiction. These tales of eagles snatching infants would appear to have their origins deep in antiquity.

Ganymede and Zeus featured on a Roman mosaic

For example, Ganymede is immortalised in Greek mythology as the man born aloft by the eagle of Zeus to become the cup bearer of the gods of Olympus. As a baby, Gilgamesh – the Babylonian equivalent of the mighty Hercules – was to be sacrificed but was instead rescued by an eagle. Such complex early myths came to be adopted, modified and distributed far and wide, even attributed to later heroes. Amongst the many myths which have come to be associated with Alexander the Great, for instance, is that he is said to have obtained knowledge of his fields of battle by first flying over them on the back of an eagle.

Etana, another mythical hero of the Babylonians, rode on the back of an eagle which fed on snakes. So, too, in Hindu tradition, for instance, Garuda, the winged mount of the god Vishnu, was an eagle giant which destroyed serpents. He had six sons from whom issued all the races of birds. In the great epic tale *Mahabharata*, Garuda issued from an egg in a flame of fire, his eyes flashed lightning and his voice was thunder – all familiar themes from eagle mythology through the ages. The Java Hawk-Eagle was chosen as the national bird of Indonesia because of its supposed resemblance to the mythic Garuda of Buddhists and Hindus.

While many such tales were first woven around the Golden Eagle, it is likely that some at least will refer to Sea Eagles. There was an ancient Act of Parliament in England:

> ... anent the slaying of the Erne on account of its being a terror to farmers, from fowls to children.

There is a story that Alfred, King of the West Saxons, no less, was hunting through the forest one day when he heard the cry of an infant from the top of a tree. He commanded one of his subjects to climb the tree where he found an eagle's nest:

> . . . and lo! therein a pretty, sweet-faced infant, wrapped up in a purple mantle, and upon each arm a bracelet of gold, a clear sign that he was born of noble parents. Whereupon the king took charge of him, and caused him to be baptised; and because he was found in a nest, he gave him the name Nestinguin, and, in after time, having most nobly educated him, he advanced him to the dignity of an earl.

Quite a tale! And one that seems not to have been unknown elsewhere. Indeed it may have provided a Stanley ancestor of the Earls of Derby with an idea. His wife had failed to provide him with a son and heir so he engineered the 'discovery' of his own illegitimate child. He was walking with his wife in Tarlescough Woods in Lancashire when they encountered a baby lying beneath an eagle's nest. The Lord of the Manor managed to persuade his rather gullible wife to accept the baby which they named Oskatel and brought up as their own. The family had even adopted the motif on their crest by the 14th century, to appear three hundred years later on some coins minted on the Isle of Man, where the family had interests. During World War I, Lord Derby enlisted some 3,000 men and was permitted by King George V to award them each silver and bronze medals bearing the Eagle and Child symbol.

The association did not end there, however, for there are still at least 26 pubs called Eagle and Child, mainly in those areas where the family had estates, such as at Gwaenysgor in Flintshire, Congleton near Allgreave, as well as Doncaster and Auckley, all in Cheshire; near Preston, Chorley, Wigan and Bispham Green, all in Lancashire; Brentwood in Essex; Staveley in the Lake District and St. Giles in Oxford. The latter was affectionately named the 'Bird and the Baby' by C.S. Lewis and J.R.R. Tolkien when, as dons at the University, they met there regularly with like-minded intellectuals. Their pub also acquired other nicknames such as 'Bird and Brat', 'Bird and Bastard' and even 'Fowl and Foetus'! It is a haunt of the literati to this day. There is also an Eagle and Child Alley in St. Giles in the Field in London.

Similar stories existed much further north, too. There is a manuscript which came to be preserved in the Advocates' Library in Edinburgh, written by one Mathew Mackaile and pertaining to Orkney in 1664:

> I was very well informed that an eagle did take up a swaddled child a month old, which the mother had laid down until she went to the back of the peat stack at Houton Head, and carried it to Choyne [Hoy?], viz four miles, which being discovered by a traveller who heard the lamentations of the mother, four men went presently thither in a boat and, knowing the eagle's nest, found the child without any prejudice done to it.

The botanist John Lightfoot told in his *Flora Scotica* (1772) how a Golden Eagle carried a child across the Minch from Harris to Skye, a distance of 20 miles. At the end of the 18th century, the *Old Statistical Account* for Strachur, in Argyll, described an eagle carrying off, and in this instance killing and devouring, a child of about three years old. The later *New Statistical Account* for Peebles recounts another tale, about a Golden Eagle:

> It had well-nigh carried off a shepherd's boy, but fortunately it was discovered hovering over its destined victim, and driven off just before it had pounced upon its prey.

The Gaelic forlklorist Charles Fergusson (1885) told of a young lad climbing into an eagle's eyrie in

Eagle and Child pub signs at Allgreave, Congleton, Cheshire (Brian Billington)

The isolated island of Tindholmur in Faroe where a mother found her baby in a Sea Eagle nest

Strathglass, Easter Ross, only to be snatched away by the eagle on its return. Unable to sustain the weight, the bird was forced to drop the boy – fortunately for him, he fell into a river so could wade ashore unharmed. In the 16th century, Martin Martin (1716) was told of a man on the Isle of Skye who was reputed to have survived a similar experience and was forever after nicknamed 'Neil Eagle'.

Fergusson pulled together other tales and observed how there was:

> … scarcely a glen in the Highlands without its story… and in Ireland I found similar stories quite common. And in both countries I was always glad to find a happy end to such exciting stories, as in every case I have come across the child was always happily recovered unhurt, generally by the heroic efforts of his mother.

One determined Faroese woman was said to have scaled the inaccessible crags of the island of Tindholmur to reach her baby, a feat never before accomplished even by the most daring of local cragsmen (Williamson 1970). In this instance, alas, the baby was dead, but its eyes were pecked out so, if in truth this incident did take place, the child must have been attacked by crows and was probably already dead when the eagle picked it up.

Happily, such fatalites seemed to be rare – and the story is so much better if the child survives their ordeal. Careful attention to detail can often make the story-teller sound even more convincing, and such embellishments would have accumulated over generations in the telling. The Shetland ornithologist Bobby Tulloch (1978) was told at considerable length of an infant from Unst, tucked behind a haystack while the family were working the field, being carried by a Sea Eagle to its eyrie on the neighbouring island of Fetlar. A rescue party set off in pursuit but when they got to the nest it was obvious that their only rope was not going to support the weight of a full-grown man. A local lad gallantly descended the cliff to rescue the baby girl who, securely wrapped in swaddling clothes, was mercifully unharmed. Many years later, she met her rescuer and the two were married, thus – to round off a good story – fulfilling a prophecy uttered by an old man as he witnessed the rescue.

Bobby was merely recording what he had been told and the detail sounded convincing enough, but this engaging tale apparently took place nearly 200 years ago. With so many 'eagle and child' stories taking place so long ago, most are impossible to validate. There might, however, be one notable exception.

Johan Willgohs (1961) was able to interview a Norwegian woman who claimed, as a child, to have been carried off by a Sea Eagle – and as recently as 1936. Again, the detail sounds convincing, except that at the time she would have been three-and-a- half years of age. It is doubtful that an eagle would ever have succeeded in carrying the child's weight of 19 kg to its nest 200 m above the ground. It is true that Sea Eagles have been known to lift weights of up to 11 kg and once a load of 15 kg, but, significantly, another had to drop a dog of this weight almost immediately. Only the largest of eagles could ever achieve these extreme feats and even then would have required the assistance of favourable winds or a downhill take-off.

When Seton Gordon (1955), our own eagle expert, attempted to interview the woman, she would only consent on the payment of a suitable fee. That in itself suggests something of a motive for claiming such a marketable story so we must doubt even her tale, which at the time probably served as a useful alibi to explain a young girl's mischievous truancy.

Seton Gordon concluded that 'no account of taking small children . . . it need hardly be said, will bear critical scrutiny,' and we must conclude that all such 'eagle and child' stories should remain firmly within the realms of folklore. With tongue in cheek – I hope – Robert Gray (1871) commented:

> It is perhaps unnecessary to congratulate ourselves that through the diligence of keepers and collectors, we are spared the infliction of seeing a modern perambulator relieved of its occupant.

Since the advent of firearms, eagles have developed a healthier respect for vulnerable and exposed situations. Indeed, one eminent professor cynically remarked that, if an infant should be borne away, it is the mother who is to blame and not the eagle.

Chapter 4
Food and feeding

A Sea Eagle exists – it has very sharp vision and while hovering in the sky, once it sees a fish in the sea plunges headlong on to it, and splitting open the water snatches it to its breast.
Pliny (AD 23–79)

It is likely that, in the quote above, Pliny was referring to the Osprey rather than to the Sea Eagle. One certainly cannot help but be full of wonder and awe when one sees an Osprey catch a fish. But the fishing ability of the White-tailed Sea Eagle (and indeed of any other sea eagle) can still be admired.

Although both species catch fish, the Osprey, with its long flexible wings, can hover as it searches for its prey. It has slit-like nostrils which close as the bird submerges and its plumage is dense and oily to improve waterproofing. The legs and feet are sturdy and strong to withstand the impact of the strike. Horny spicules cover the soles of its feet and, together with the long sharp curving talons, facilitate the grasping of slippery fish. The latter features, though less developed, are also to be found in the White-tailed Sea Eagle. I have also noticed Sea Eagles reverse the outer toe to hold a fish, so that two toes are held forward and two back, but this is accomplished without the facility of the Osprey. Rarely does the Sea Eagle plunge-dive under water, preferring instead to snatch fish lying close to the surface or in shallow water; thus the spectrum of prey species available is more restricted. The Sea Eagle is, of course, a larger and more cumbersome bird. Its powerful beak is a conspicuous feature, used to pierce and tear with a strong twisting and pulling action. Yet it is wonderful to witness the dexterity and finesse with which a parent eagle will employ this formidable tool to present even the tiniest morsel of food to a newly-hatched chick.

Like the Osprey and all other raptors, sea eagles possess a highly-developed sense of sight. To anyone, the 'eagle eye' instantly sums up the epitome of visual acuity. William Macgillivray (1886) commented:

> Floating hundreds of feet above the summit of our highest mountains, the eagle spies a grouse or hare – where you, my honoured reader, would hardly discern an elephant.

An eagle is said to be able to detect a rabbit a kilometre away, provided that the wee beast is running. Some authorities claim that an eagle's vision is eight times better than ours, while others are more conservative, putting it at only two or three times. Whatever the answer, the result is impressive, thanks to several unique adaptations. Apparently, in the eye of a raptor, the sensory cells tend to be concentrated in the upper hemisphere of the retina so that, as an eagle flies, images from below can be perceived more clearly. When perched and wishing to engage an object on the same plane, however, an eagle might risk all dignity by turning its head almost completely upside down, focussing with a penetrating but distinctly quizzical stare. Many raptors such as eagles are unusual in possessing two foveas; we have only one. (The fovea is a pit in the retina with a particularly high density of receptors, or cones, providing a keen perception of movement and distance.) Finally, oxygen is fed to a bird's retina not via blood vessels, as is the case in humans and most other mammals and which can reduce visual acuity, but by means of a complex flap of folded tissue called the pecten, a feature that is especially well-developed in raptors.

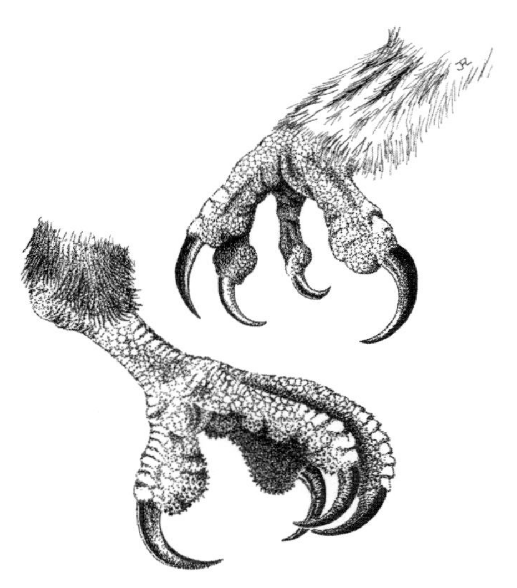

The talons of a Sea Eagle (left) and Golden Eagle

An eagle has only a 40° overlap in its field of vision compared with 120° in humans. The eye is protected under a bony projection and by a third, nictitating, membrane, which is translucent and flicks across to clean and moisten the lens. It is likely that the Sea Eagle will deploy its third eyelid just before it strikes at prey, just as a gannet and other birds do. In all birds the eyes are large in relation to the head, and those of the Sea Eagle – at 46 mm in diameter – are larger than those of a human (24 mm), and almost as large as an ostrich which is eighteen times heavier (Birkhead 2008). Relative to body size, raptors such as eagles and owls boast the largest eyes in the avian kingdom. This leaves little room in the orbit for muscles so both eye movement and distance perception have to be achieved by moving the head on a flexible neck.

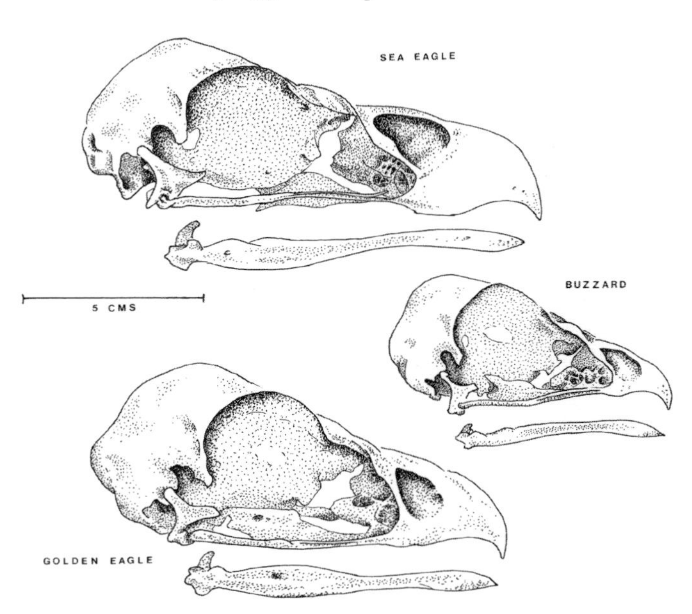

The skulls of a Sea Eagle (top), a Golden Eagle (bottom) and a buzzard (middle)

Although sight is of prime importance in the location and pursuit of prey, the Sea Eagle can be attracted to carrion by the mere presence of other eagles, corvids, gulls or vultures. There is no evidence that scent is involved, though all birds may possess some limited power of smell. It has been shown, for instance, that New World Vultures utilise smell to detect carcasses.

As the eagle soars in its search for food, the long, broad wings 'fingered' at their tops, it presents its familiar silhouette. With air turbulence – and hence drag – most pronounced at the wing tips, these deeply-notched and emarginated primaries function as aerofoils; to reduce drag, effectively lengthening the wings and enhancing lift. The total wing span of a White-tailed Sea Eagle may exceed 2.5 m.

Wing spread of adult Sea Eagle

The crop or oesophagus of birds has an elastic wall and that of the Sea Eagle can easily hold several days' sustenance, permitting it to exploit temporary abundances of food. It can therefore afford prolonged stints loafing on a favourite perch or soaring effortlessly in air thermals. A brief radio-telemetry trial I undertook in Rum in 1977 (Love 1979) indicated that one female spent only 8 per cent of its time in the air, although during the 4-day observation period the weather was bad and may have grounded the bird more than normal; because the bird was newly-liberated it was also probably somewhat reluctant to venture far. Nonetheless the results were similar to more detailed observational studies by Leslie Brown (1980) on the closely- related African Fish Eagle which only spent some 10–25 per cent of its day active.

Extended periods of inactivity, so typical of predators, together with large bulk in relation to body surface, serve to minimise the Sea Eagle's food requirements. Like many predators, Sea Eagles spend a lot of time loafing around and favourite perches can often be quite obvious. They accumulate droppings and food remains to form a mound or hummock of vegetation which the Norwegians term 'ørnetue'. These are obvious places to seek evidence of the Sea Eagle diet.

So what do Sea Eagles eat? Pellets collected in quantity from beneath a nest or favoured roost can be useful in investigating the diet of wild raptors. However, biases can be introduced.

An eagle can digest some large fish bones so they may rarely appear in pellets. On the other hand, feathers, fur and animal bones are the stuff of most pellets and can readily be identified. Not every prey species contributes to a pellet with the same degree of facility, though. Fur-bearing prey, especially lagomorphs and squirrels, tend to be detected more easily, whereas game birds, which are relatively loose-feathered and easily plucked, are underestimated in pellets. Carrion tends to leave bone chips and encourages the appearance of grass in pellets; afterbirths leave no trace at all in pellets.

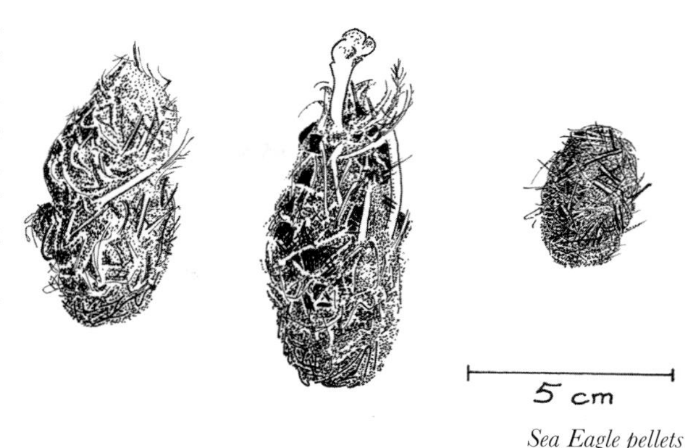

Sea Eagle pellets

Studies of Sea Eagles in Greenland by Danish ornithologists have further elucidated problems in data collection (Wille 1979). Small fragments of prey and pellets collected from nests and brought back to the laboratory for identification showed a distinct bias towards bird prey (68 per cent of items) at the expense of fish (20 per cent). In contrast, more complete items lying in a nest and readily identifiable in *situ* comprised 58 per cent fish and only 30 per cent birds. Prey remains collected from 40 eyries in the same vicinity – as might be expected – were found to consist of 68 per cent fish and 16 per cent birds. A completely different pattern emerged, however, when automatic cameras were mounted at five different nests to photograph each food item carried in by the adults. Fish was found to constitute 93 per cent of the diet and birds only 5 per cent, reflecting the true diet of Sea Eagle pairs in that part of Greenland.

The use of such sophisticated data-recording equipment is both expensive and time-consuming. Normally one has to rely on the simple identification of prey remains, all the while bearing in mind the limitations, for example, that fish will be underestimated and birds overestimated. While such a basic technique will not reveal the precise diet of the predator in question, it can be used to compare diets from different habitats. Sea Eagle prey lists now exist from several countries. As we would now expect, most show a preponderance of bird prey (39–87 per cent of items), although one exception revealed 79 per cent fish from the middle reaches of the River Danube where bird prey was in short supply (Fischer 1979). Only inland on the Kolsky Peninsula of northern USSR (Flerov 1970) did mammal prey assume any importance (41 per cent of 172 prey items) where birds, and possibly fish, too, might be less abundant than on the coast. Sea Eagle diets may vary within a relatively restricted

Adult Sea Eagle bringing food to its nestlings, Greenland (F. Wille)

area. In Greenland, fish constituted up to 50 per cent of the items taken well up the fjords, while on the outer coast and skerries the numerous seabird colonies provided an alternative abundance of prey (Kampp and Wille 1979; Wille 1979). Even individual pairs of eagles may display dietary preferences, dependent upon a particularly favoured hunting technique and what is most readily available within the hunting range.

It should be borne in mind that prey found in a nest may not reflect what the adults themselves consume. Fish with easily digestible bones might be given preferentially to eaglets, since their growing bones and feathers will demand an enhanced calcium intake. Meat is rich in phosphorus and therefore less desirable at this time since it can seriously disrupt the ratio of calcium to phosphorus within the eaglet's body; studies on vultures feeding exclusively upon carrion have demonstrated how bone fragments in the meat regurgitated by the parents are essential to the chick's normal development.

It is during the winter months that White-tailed Sea Eagles, both adults and immatures, seem most liable to utilise carrion, as several early Scottish accounts illustrated, including Dixon (1900):

> They are also very unclean feeders, being little better than vultures in this respect. Of the two the White-tailed Eagle is the worst; he is a regular scavenger of the shore and in not a few cases we have known him lured to his doom with a mass of stinking offal, a putrid lamb or decaying fish. Healthy vigorous birds or animals are seldom attacked by this Eagle; it confines its attentions to the weakly and the wounded creatures that cannot move fast or offer any serious resistance.

The diet will differ in winter time when a different spectrum of prey becomes available. Amongst winter observations from 19th-century Scotland that have come to my attention, 70 per cent were of mammal prey. Since these reports were gleaned from the literature, however, they may not be truly representative. Nevertheless, prey collections made in Norway and in East Germany tended to confirm how mammal prey became more favoured in winter (bearing in mind, too, that carrion will tend to be underestimated among the food remains used in the analysis). The former study (Willgohs 1961) also revealed how birds – auks rather than shag, eider or gulls – became a more important constituent of the winter diet when fish such as cod, lumpsucker and wolf-fish had moved to deeper water. Being opportunists, the eagles could procure some fish by other means too, as carrion or as waste from humans, otters and gulls, for example. In Baltic Sweden, birds became more important than fish in autumn and winter, while further north in Lapland, reindeer carrion assumed some importance (Helander 1975).

An immature Sea Eagle feeding on carrion, Rum

As one accumulates a list of prey species, it becomes increasingly obvious just how catholic Sea Eagles are in their tastes. A cursory survey of the readily accessible literature reveals about 80 species of birds and over 30 species of fish, and a similar assortment of mammal species have been recorded. Not surprisingly some species feature more prominently than others, depending upon the hunting technique employed.

Fish

The Sea Eagle most commonly hunts by gliding a few metres above the water, sometimes pausing momentarily before descending to strike. At the last moment the feet are brought forward to snatch the prey with barely a splash. When searching from a height of 200 m or more, the sudden swift stoop can terminate in a full immersion similar to that of the Osprey, but Willgohs (1961) asserted that plunge-diving is a rare occurrence. It is not unusual for a hungry Sea Eagle to operate from a perch overlooking water, such as a convenient crag, hillock or tree. On the northern coasts of Norway high tussocks on low, grassy islets are used. Repeated defecations by the easgles encourage grass growth so that these traditional perches – termed 'ørnetuer' – grow bigger each year. A Sea Eagle may sit on the shore or even wade into the shallows to seek fish, as would a Heron. Willgohs also recorded an unusual instance of two eagles sitting on open water, apparently attracted by a shoal of small fish; both birds took off without difficulty.

Lumpsuckers: the male is smaller (Harald Misund)

If the Sea Eagle lacks the finesse of the Osprey, it procures deep-swimming fish by other means, not least by blatant piracy, notably from the Osprey itself. It readily steals fish landed by Great black-backed Gulls and Herring Gulls, especially lumpsuckers, from which the gulls, characteristically, had eaten only the eyes and the guts. Food may also be stolen from Snowy Owls, Red Kites, Black Kites, Buzzard and Peregrines. I have also watched Ravens and Hooded Crows being deprived of food. As long ago as 1774 in Orkney (Low 1813) Sea Eagles were mentioned as lifting fish from otters. Both Johan Willgohs and Harald Misund affirm how otters contribute much to the diet of Sea Eagles. Indeed, Norwegian fur hunters often learnt of the whereabouts of otters by spotting an eagle glide down and deprive it of its fish.

Nor is the Sea Eagle averse to scrounging fish around harbours, fish markets or from fishing boats. As many as eight eagles have been seen to join the gulls following boats while the crew guts the catch. Willgohs mentioned an unusual instance of a Sea Eagle found dead in a trout net, from which it had presumably attempted to steal. Accounts mention how 'the species often made considerable havoc in carp ponds', while in the Sava valley of the former Yugoslavia, a pair fed their eight- to nine-week-old-brood almost entirely on decapitated carp. There have been no complaints from Scottish fish farms, but a mussel farm in Mull quite appreciated the presence of Sea Eagles since they kept the Eiders away.

Thus, by various means, a wide variety can be taken by Sea Eagles. Willgohs (1961) listed 328 fish of 24 species taken in Norway; nearly 50 per cent of the sample comprised only two species – lumpsucker *Cyclopterus lumpus* (24 per cent) and wolf-fish *Anarhichus lupus* (17

Young Sea Eagle 'swimming' ashore, Norway (Harald Misund)

per cent). Cod *Gadus morhua* and other gadids, such as pollack, saithe, haddock and whiting, constituted a further 27 per cent with sea perch *Sebastes marinus* another 10 per cent. A shorter list of 91 specimens of nine species from Greenland (Kampp and Wille 1979) found 56 per cent made up of gadids and 33 per cent char *Salvelinus alpinus*; among the remainder were five scorpion fish *Cottus scorpius* and single specimens of *Cyclopterus* and *Anarhichus*. Seventeen gadids (52 per cent) featured among 33 fish from the White Sea (Flerov 1970), with seven each of *Anarhichus* and flatfish *Pleuronectes* (each 21 per cent) and two herring *Clupea harengus*. Wolf-fish and lumpsuckers appear also to be common prey in Iceland. A dozen cod were found in one eyrie in the Isle of Skye. *Raia batis* and halibut *Hippoglossus* have been recorded in Shetland eyries in the past. Gurnards, skate and halibut were commonly featured since they frequently bask on the surface. The Shetland naturalist Thomas Edmondston (1844) observed:

> Fish, I believe, is his general food; and he boldly attacks the largest kinds if they happen to come to the surface. Several desperate combats have been witnessed between this bird and the halibut. The former strikes his claws into the fish with all his force, determined not to forego his hold, and although but rarely, is sometimes drowned in the attempt to carry off his prize. When he has overcome the halibut, he raises one of his wings which serves as a sail and if favoured by the wind, in that attitude he drifts towards the land. The moment he touches the shore, he begins to eat out and disengage his claws, but if discovered before this can be affected, he falls an easy prey to the first assailant.

Charles Fergusson (1885) recounted the tale of a servant girl on a small island in the Sound of Harris, near Berneray. She saw what appeared to be a small sail boat drifting ashore:

> She could not conceive what kind of craft it was till it touched the shore, when to her astonishment, she found it to be the dead carcase of a cow and the sail the spread wings of an eagle, that had its talons so deeply embedded in the carcase that it was utterly unable to extricate itself or escape. The girl unfixed its talons and took hold of the bird, but no sooner had she done this than the ferocious bird fixed its talons in the girl's thigh and tore out the flesh from the bone. The wound healed up, but a hollow large enough to hold an

> apple remained as the effect of the injury. The woman lived to an advanced age, and was an ancestor of the Captain Malcolm Macleod, who told me this.

Frank Wille has photographed a Greenland Sea Eagle 'rowing' ashore with its wings whilst still clutching a fish which was too large to lift from the water. Indeed it has been claimed that eagles can drown in this manner and the number of such cases documented (Willgohs found 35 in Norway alone) makes it difficult to discount them. Most involved halibut or large salmon *Salmo salar*. In one or two accounts the witness claimed to have seen the eagle sucked under and drowned, while another even heard a loud crack – the bird's wings breaking as it was dragged beneath the surface. A large halibut has even been found with an eagle's feet still embedded in its back, the bird having rotted off. Ospreys may similarly succumb, although it should be mentioned that a photograph was published in the journal *British Birds* which purported to show the skeleton of an Osprey with talons firmly embedded in the body of a carp; Cowles (1969) from the British Museum later demonstrated this to have been a hoax. Not only had the bird been attached artificially after the fish had died but the skeleton in fact belonged to a Buzzard.

Whether all these accounts and observations are accepted or not, they do raise the question of how heavy a fish a White-tailed Sea Eagle can actually carry. Willgohs quoted a claim that an eagle had lifted a halibut weighing 15 kg, but described an instance of another eagle which had dropped a cod weighing only 8 kg. It would seem, however, that most fish caught weigh from 0.5 to 3 kg and that the smallest are herring and scorpion fish, measuring 15–20 cm long. Smaller fish feature rarely in the diet.

In fresh water, a completely different range of species is available to the Sea Eagle and by far the most commonly caught fish is the pike *Esox lucius*, a species much given to swimming immediately beneath the surface. It seems that specimens weighing 2–5 kg, with lengths of 60–90 cm, are those most frequently taken. Sixty pike were taken by Ernes on the Muritz river in East Germany (Fischer 1970), together with three bream *Abramis brama*, a perch *Perca fluviatalis* and an eel *Anguilla anguilla*. A further 36 fish taken from eyries on the Darss Peninsula (Schnurre 1956), also in East Germany, included eight pike, 14 eels, seven cod (taken offshore) while perch, carp, trout, shad *Alosa alosa* and roach *Rutillis rutilis* have also featured as prey. In the Quarken Straits of Finland (Stjernberg 1981), pike constituted 72 per cent of the fish diet, with perch, roach, bream and orfe *Leuciscus idus* making up the remainder. In Ireland, Sea Eagles were often seen watching over fords where salmon leapt, although dead or spent fish were also taken (Ussher and Warren 1900). Willgohs aptly concluded that:

> The eagle will probably take almost any fish species of convenient size where it can easily be obtained, and therefore the fish diet may vary considerably from place to place and from time to time.

Adult Sea Eagle with fish, Isle of Skye (Bob McMillan)

But the last word on this fishy tale must surely go to a fisherman:

> The angler... must regret [the Sea Eagle's] disappearance for it would be a pleasant interlude in his operations to watch the doings of these lordly birds, and give a further wildness to the scene to hear their cries echoing among the rocks.

Birds

As with fish, although some 80 bird species have been recorded, only a few tend to be favoured. Again Sea Eagles can employ diverse techniques in their capture.

Colonel Richard Meinertzshagen (1878–1967) watched an eagle being violently mobbed by gulls, until with a sudden and elegant twist, it succeeded in grasping one reckless individual which it then took to a nearby post to pluck and eat. Amazingly, a Goshawk has been seen to succumb to a Sea Eagle in like manner (Willgohs 1961) but on such occasions the eagle is perhaps displaying more good luck than judgement. Reports have been made of a Heron caught in flight and a Mallard struck in mid-air, whilst aerial attacks on larger geese were often unsuccessful (Rudebeck 1951).

An element of surprise, together with a tactical use of cover or bright sunlight, may enhance an eagle's chance of success. Willgohs noted an eagle passing low over the water to snatch an unsuspecting heron on the shore, and I have seen a Sea Eagle in Rum flying low over a sand dune with bright sunlight behind, attempting to surprise a flock of gulls on the shore.

Moulting birds, such as flightless ducks and geese, can be caught with relative ease. Like most predators, the Sea Eagle is also ready to capitalise upon any debilitated animal. Ducks injured by hunters are easy pickings while both Mute and Whooper Swans frozen into the ice in lakes have been taken in Germany. An injured Curlew has been attacked, while an injured Bean Goose successfully defended itself with its wing (Rudebeck 1951). The deliberate selection of debilitated prey is interesting in itself, demonstrated by observations on Sparrowhawks, Merlin and Peregrine where Rudebeck was able to recognise injured or otherwise abnormal individuals amongst about 20 per cent of the prey taken. By being selective to increase its chances of success, a predator is performing a useful 'sanitary' function on the prey populations. In this particular study, only three successful Sea Eagle hunts were witnessed, and one victim of these was a Mallard unable to fly. The total sample included a total of 60 'hunts', so only 5 per cent proved successful. This seems exceedingly low and may result from including many low-intensity stoops which even the eagle itself may not expect to yield results.

Adult Sea Eagle chasing a Great black-backed Gull, Skye (Bob McMillan)

Sea Eagles released on Fair Isle and Rum soon learnt to catch both gulls and Fulmars in flight. Fulmar chicks might be approached as they sit on the nest ledge, but the eagle risks being spat upon. And indeed, as we have seen already, this foul, oily liquid proved the ultimate undoing of one of the Fair Isle eagles. In Norway, Canna and Rum, Sea Eagles have visited seabird colonies to lift young Kittiwakes from nests. Heronries can be raided, as can the nests of Rooks, Black-headed Gulls and Kittiwakes, while even the nestlings of Ospreys, hawks and Black Kites might be taken. Young birds are, of course, especially vulnerable and 86 per cent of the Black-headed Gulls killed on the East German lakes were juveniles. John Wolley (1902) noted 'heaps of young herring gull remains' at an eyrie on Dunnet Head in Scotland. In Norway, Great black-backed Gulls were seen snatching young seabirds on their maiden flight to the sea, then in turn being pursued by piratical Sea Eagles (Willgohs 1961).

Exceptionally, Sea Eagles may take eggs which, according to one authority, may be carried in the beak. In Norway it is suspected that the eggs of Kittiwakes, Eider, Shag and gulls have been eaten (Willgohs 1961). Incubating Eider can be killed and I recorded two such instances on Rum. Eider ducks flushed from a nest fly ponderously and perhaps weakly, so they are easy prey.

Typically, however, a Sea Eagle hunts low over a flock of birds on the water, repeatedly swooping to attack one luckless individual whose only escape is to dive again and again until, exhausted, it emerges for its final breath. The Russian ornithologist Flerov (1970) once watched an eagle make seven such attacks on an Eider, and 12 on another. In neither instance could the eagle lift its prey from the water, but instead dragged it to the shore. Up to 65 attacks have been recorded in periods of 35–45 minutes. Not surprisingly, such efforts can exhaust the eagle too, and one immature had to give up after 15 and 28 attempts at Dabchick.

After an unsuccessful stoop at a diving bird, the eagle circles low to keep close to the bird when it next emerges. It is easier to follow the prey beneath the surface in shallower water, and Willgohs suggested that male Eiders more frequently fall victim because they are more conspicuous under water. A pair of eagles sometimes proves to be a more effective team, one being near the surface when the victim emerges, the other remaining at a suitable height to watch the movements of the prey under water and to assume position for the next attack. The Swedish artist Bruno Liljefors (1860–1939) vividly painted two Sea Eagles attacking a Red-throated Diver, a scene strikingly similar to one painted by J.G.Millais, who claims to have witnessed two eagles tiring out a Great Northern Diver near Lofoten. In his book *A Reed Shaken by the Wind*, Gavin Maxwell related how he watched several thousand Coot bunching together in panic on a Euphrates marsh, attempting to evade repeated attacks from five Sea Eagles.

Carrying a coot, for example, an eagle might be able to tuck up its legs, but with larger prey its legs hang down. Both feet may be required to carry larger prey but only one for small prey. Several times Sea Eagles have been seen carrying one eider chick in each foot.

Repeated harrying of birds on the water is most successful with diving birds, and consequently such species figure prominently in the diet. Eider constituted 30 per cent of avian prey items in Norway and around the White Sea in the USSR. During the winter months in Finland, Sea Eagles were attracted by the huge concentrations of Long-tailed Duck and Goldeneye. Other species occurring in the Norwegian sample were Shags (10 per cent), guillemots and other auks (28 per cent), with gulls forming another 13 per cent. Eider and Glaucous Gulls were common prey in Iceland with some Fulmars, Ptarmigan and an occasional merganser. In Greenland (Kampp and Wille 1979) Eider were again favoured (20 per cent of avian prey) and to a lesser extent Long-tailed Ducks (6 per cent); auks would seem to have been uncommon in the area so Iceland and Common Gulls amounted to 26 per

Adult Sea Eagle

cent of the prey items, Ptarmigan (20 per cent), Ravens (13 per cent) and Red-throated Divers (8 per cent) all featuring instead. Landbirds play a minor role in the diet of Norwegian Sea Eagles but in the White Sea a variety of grouse and waders, especially Snipe, were taken.

In the Quarken Straits area of Finland (Stjernberg 1981), Red-breasted Mergansers, Goosander, Great Crested Grebe, Black-headed Gull, Mallard, Goldeneye, Velvet Scoter and Tufted Duck/Scaup were taken. On freshwater lakes, diving birds again featured prominently. In East Germany, Coot were important – 16 per cent of avian prey in one study (Flerov 1970) and 70 per cent in another (Schnurre 1956). Mallard made up 19 per cent and 16 per cent in these respective samples. Dabbling ducks are less prone to capture by Sea Eagles because their alarm reaction is to leap from the water and take flight immediately. Individuals can sometimes be snatched from dense, panic-stricken flocks, but commonly only isolated birds are attacked. Disadvantaged birds are particularly vulnerable, especially those injured, in moult or 'pelleted' birds which narrowly escaped the barrel of wildfowlers. Gosling Greylags (23 per cent) and juvenile Black-headed Gulls (25 per cent) are easy prey (Fischer 1970), together with diving species such as Coot and grebe.

Willgohs (1961) was told of an eagle which pushed into a dense flock of Starlings to emerge with one in its talons. This and a thrush are the smallest recorded prey taken by Sea Eagles, but are taken infrequently. Such small items may be under-recorded, however, since they leave fewer conspicuous remains. In any case, for the effort involved in catching them, they reward an eagle with little food value. Species such as sandpipers and Turnstone are killed rarely and most bird prey range from 1.5 to 2.5 kg in weight. The largest are swans, up to 10 kg, and Great Bustard, at about 15 kg.

A similar picture emerged when I searched old Scottish records. Prey species included gulls (mainly Herring, Great black-backs and Kittiwakes), Guillemots and Puffins. John Wolley (1902) mentioned 'thousands of Cormorants' being found in one eyrie, although not only was this a gross exaggeration, but he probably really meant Shags. As in Norway, gamebirds were taken rarely and Booth (1881–87) admitted that in this respect the Sea Eagle was less destructive than the Golden Eagle. On Skye, however, Captain Macdonald recorded several grouse being taken by Sea Eagles, one being chased right across a loch before managing to find shelter in a hole! And, of course, Captain Macdonald's captive Sea Eagle would retrieve grouse that its master had shot. Domestic fowl can be easy prey, however, and there was mention of several onslaughts on the farmyard by Sea Eagles, apparently causing much anxiety among the local inhabitants.

Other prey

Before going on to consider mammalian prey and domestic stock in particular, brief mention should be made of a few other items that have appeared on the Sea Eagle's dinner plate. One of the East Scotland releases was apparently seen hunting frogs; they and small reptiles have been recorded being taken elsewhere, including an adder *Vipera beris*, which was immediately rejected. Captive Sea Eagles in Israel were seen to catch and eat small swamp turtles *Clemmys caspica* which frequented the ponds in their aviary (Edna Gorney, pers. comm.). A Sea Eagle was observed seeking molluscs and snails on the banks of the Rhine, while another was seen robbing crows of freshwater mussels *Anodonta* and dropping them to break them open. Interestingly, the eagle itself then dropped any that remained unopened, mimicking the crows ineptly by being only a metre or so from the ground. Some reports suggest that cuttlefish, crabs, lobsters, starfish and sea urchins have been taken. One Sea Eagle is said to have gorged on jellyfish *Aurelia* or *Cyanea* but Willgohs thought that this must have been a diseased or injured eagle. Miscellaneous shells of marine mussels *Mytilus* and snails *Gibbula*, *Buccinium*, *Littorina* and the like, occasionally found in nests, at roosts or in pellets, most likely originate from the stomachs of eiders which the eagles have eaten.

Mammals

During years of population explosions of voles or lemmings in the Arctic, these rodents may be caught by Sea Eagles. Indeed, half of the mammal prey species in the White Sea area were voles (Flerov 1970). Willgohs recorded an occasional water vole *Arvicola terrestris* in pellets collected from the outer skerries of Norway, and this seems to be commoner prey further north. Marmots *Marmota*, susliks *Citellus* and mole rats *Spalax* may be taken as prey in the Russian steppes. Rats *Rattus norvegicus*, field mouse *Apodemus* and other mice, hamsters, moles *Talpa europea*, hedgehogs *Erinaceus europeus* and squirrels *Scurius vulgaris* have all been found at eyries. Muskrats *Ondatra zibethica* and mink *Mustela vison* featured to a minor extent in the diet of White-tailed Sea Eagles in the Quarken area of Finland (Stjernberg 1981). A mink cub has recently been found in a Scottish eyrie. On the Ili delta of Kazakhstan, muskrats from local fur farms were taken in the spring and autumn, when they constituted 30 per cent and 43 per

Adult Sea Eagle stooping on prey, Skye (Bob McMillan)

cent of all items found at Sea Eagle eyries. For some reason they were less favoured – or were simply less available – in summer (14 per cent of items) when voles might be caught instead (Glutz von Blotzheim et al. 1971).

Probably the most common live mammal prey items taken by Sea Eagles are rabbits *Oryctolagus cunniculus* and hares. Arctic hares *Lepus timidus*, with their white winter coat, seem to be especially vulnerable when there is little snow or where cover is sparse, as on offshore islands in northern Norway; in Greenland, the equivalent *Alopex lagopus* appears as prey of Sea Eagles. It is not unknown for two or three birds to operate together to catch hares. In Britain, brown hares *Lepus europaeus* and rabbits were common winter prey, while both common and Arctic hares were occasionally taken during the breeding season.

William Macgillivray (1886) asserted that Sea Eagles were 'especially fond of dogs'. Willgohs recorded how one unfortunate hound of 11 kg was carried off but that another of 15 kg was too heavy and had to be dropped. Cats may at times also fall victim, together with a variety of small carnivores such as martens *Martes*. It would seem that otters *Lutra lutra* are rarely attacked, although Willgohs mentioned an instance in Norway, and another in Sweden, where a bitch otter was killed just as it gave birth to its two young (Uttendörfer 1939). I am surprised that the mother was doing this in the open and not in the safety of its holt. There is an old account of a fight between an eagle and an otter on a loch in southern Scotland. Dixon (1916) did not think they were contesting a fish. A second eagle appeared on the scene and managed to sink its talons into the otter before it finally escaped. A live jackal *Canis aureus* which had been caught in a trap was taken by a Sea Eagle, while a silver fox was killed after escaping from a fur farm in Norway. Two foxes are said to have been lifted by eagles during organised 'fox drives' in Norway, doubtless much to the annoyance of the hunters. There are also the inevitable tales of furious battles between eagle and fox, and in Norway a fox skeleton has been found with the bones of an eagle firmly attached.

Red deer calf, Isle of Rum

An eagle was once seen to lay hold of a sleeping seal which instantly dived, taking the bird with it; shortly afterwards the eagle reappeared with its wings broken and on the point of death (Bannerman 1956). Attacks on young seals are not uncommon, Seton Gordon (1955) pointing out how easily a sleeping seal might be mistaken for a dead one. Willgohs even documented a Sea Eagle drowning in an attempt to catch a live porpoise *Phocoena phocoena*. In the main, however, seals and cetaceans feature in the eagles' diet only as carrion. The same may be said for most ruminants, although calves of roe deer *Capreolus capreolus* or reindeer *Rangifer tarandus* can be taken alive. Fiona Guinness (pers. comm.) watched a young newly-released Sea Eagle in Rum kill a new-born red deer calf *Cervus elaphus*. There are only one or two records of full-grown deer being killed. One panic-stricken roe was seen to collapse exhausted with an eagle firmly clinging to its back (Willgohs 1961) but another, more fortunate beast on the Isle of Harris successfully evaded capture (Macgillivray 1886).

Predation on domestic stock

The periodic appearance of domestic stock on the diet sheet of the Sea Eagle has made the bird many human enemies. Piglets have on rare occasions been lifted, while C. M. McVean countered that his tame eagle was afraid of pigs (Gray 1871). Goats have also been attacked and indeed the carcass of almost any animal, even a cow, may be an attraction to a hungry eagle. At certain times of the year there may be an abundance of available carrion. In Argyll, on the west coast of Scotland, for instance, around 6–7 per cent of blackface ewes die each year; not all are buried, as is demanded by law. It is not surprising, then, that in 1959, along a 2-mile transect in Lewis, Jim Lockie and David Stephen (1959) encountered no fewer than 28 dead sheep. Overstocking in the past may have contributed to a gradual deterioration in grazing quality, as it did last century in Rum and elsewhere. As Booth attested:

> Mutton, I think, in one form or another, is the usual diet of this species, and for lambs it has without doubt a great partiality. It nevertheless exhibits a cowardly nature in procuring even such a helpless prey, as it is seldom that an Eagle will swoop down and carry off a lamb that is not separated from its mother. When, after a protracted winter and a dry cold spring the herbage is scarce, so many dead carcasses may be found in all directions, scattered over the moors.

In early May 1849 John Wolley met a gentleman in Assynt who lamented how Sea Eagles:

> ...were very numerous, for he had seen seven together not very long before. He said there was one which was very destructive to his lambs. I was able to tell him I had probably shot that bird. The next day I saw a shepherd who had himself killed thirty Eagles while the rewards were given. When once Eagles begin to kill lambs they continue to do so...they are only said to take them on a windy day, when they can rise easily with a weight. Several people told me this.

Immature Sea Eagle feeding on carrion, Rum

Somewhat surprisingly, perhaps, that eagle was the only one shot by Wolley in his albeit rather short life. He sent it to England to be stuffed and it 'is now in the possession of Mr Edge at Shelley, where I saw it in 1856'. Wolley died four years later, aged only 37.

Lambs are especially vulnerable, and with the ready availability of dead or dying ones, it is understandable that an eagle may also occasionally be tempted to lift one that is still alive. When an unfortunate shepherd witnesses his livelihood disappearing in this way, one can appreciate his animosity. It may be only a few 'rogue' eagles which bring upon their innocent peers an ill reputation, but it can rapidly generate a determined campaign of destruction. Rev. H.A.Macpherson (1892) expressed a typical sentiment, based on experience in the Lake District but one that could be felt anywhere that eagles were to be encountered:

> ...these birds were so destructive to the interests of the shepherds, that their extermination became absolutely necessary.

In summary, the conclusions of Robert Gray (1871) seem appropriate:

> The chief food of this eagle appears to be stranded fish, procured in the vicinity of its maritime haunts, dead sheep found on the moors, and occasionally a salmon left by some scared otter – a selection more in keeping with the innocent life of a vulture than the plundering habits usually ascribed to eagles.

To this can be added the Sea Eagle's ability to catch fish: mammals are rarely an important food and mostly taken as carrion. Sheep and lambs have generated more emotion among shepherds than nourishment for eagles. As we shall see in the final chapter, the evidence would indicate that extermination of the Sea Eagle because of its depredations upon domestic stock was totally unwarranted – a deliberate and reprehensible act of vandalism.

Chapter 5
Sea Eagles in Britain

The White-tailed Eagle usually chooses for its retreat some lofty precipice overhanging the sea, and there in fancied security forms its nest and reposes at night.
William Macgillivray (1886)

One summer evening in 1958, crofter Ronnie Simison was walking over his farm at the southern tip of South Ronaldsay in the Orkney Islands. He paused to contemplate a mound on the cliff top. When he dug down a little, he uncovered some polished stone axe-heads, a blade and a button of black jet. After showing them to Stromness Museum he returned to expose more of the wall, and some bones, before breaking into an open chamber which, when illuminated with a cigarette lighter, was seen to be littered with human skulls.

Surprisingly, archaeologists registered only mild interest and it would be two decades before they decided to excavate on Ronnie's property, digging not at the tomb but on a nearby burnt mound. When his farm work allowed, Ronnie helped and, through looking, listening and learning acquired some skills of his own in excavation. It was not long before he could resist it no more, and in the summer of 1976, he reopened his burial chamber, carefully recording his progress and logging all his finds, including thousands of bones, most of them human.

Ronnie's friend and neighbour, the archaeologist John Hedges, took up the story in a book published in 1984:

> Among the [animal bones] Ronnie Simison could identify some of sheep and cattle but it was the bird bones that aroused the strongest interest. At first he recognised talons – substantial numbers of them – and then he found a large skull which was sufficiently complete to leave no doubt of the bird in question. Astonishingly, the human and animal bones in all three contexts had those of eagles mixed among them. It was a dramatic discovery and one which was to prove to be of considerable importance... Very few [of the bird bones] were from species which have been taken for food, a staggering 97 per cent coming not just from birds of prey but, for the most part, from ones that will eat carrion! The most singular fact of all is that 90 per cent of the bones of this type of bird belonged to one particular species – the magnificent white-tailed sea eagle... I would suggest that something like 35 carcasses of birds of prey had been involved in total and that two-thirds of these were white-tailed sea eagles.

Along with a barrowload of pottery fragments and other artefacts, Ronnie and the archaeologists eventually uncovered a staggering 16,000 bones, representing about 340 people. Among 745 identifiable bird bones there were no fewer than 641 eagle bones (including 70 talons) but later conservative estimates reckoned they represented only 8–10 individuals. Ravens, another scavenger, were also present. Originally interpreted as a foundation deposit, laid just as the tomb was built, new carbon-dating techniques revealed that, although the tomb was begun around 3,000 BC, the eagles died from 2,450–2,050 BC, up to 1,000 years later. Obviously the tomb had remained in use for many generations.

Sea Eagle skull and talons excavated at the Neolithic Tomb of the Eagles, Isbister, Orkney

A recent re-examination of 85 human skulls from the Isbister Neolithic tomb showed how at least sixteen – both men and women, even children – exhibited some serious, perhaps fatal, head wounds caused by a stone or a mace, even an axe. Some of the victims had recovered, the skull of one woman having three healed wounds from blows by a blunt object. She also had a dislocated jaw which was badly healed so, although living to tell the tale, she must still have suffered terribly. Most of the bones seem to date from the fourth millennium BC and seem to have been the result of deliberate attack rather than ritual killing. It seems that the life of Neolithic farmers had not been quite as peaceable as previously assumed.

Carrion-eating birds like eagles and ravens were obviously held in some reverence at Isbister. As we will discover, eagles and ravens would often frequent battlefields to feed on the bodies of the slain, so I offer that as one possible scenario. John Hedges, though, had noticed how the eagle bones tended to be quite white compared with the stained human bones. He postulated that the human bodies might first have been laid out in the open to be stripped of flesh by scavenging birds, perhaps on elevated platforms as used, for example, by the Dakota and Choctaw tribes on the North American plains. Such 'sky burial' practises are known from other cultures, notably Parsees and Iranis in India who depend upon hoards of city vultures to feed in specially constructed stone 'Towers of Silence'. Mongols similarly abandon corpses in high places for wolves and eagles to consume, and there is a saying amongst the Chinese of Central Asia that the eagle is a nomad's coffin.

At Isbister it seems, as is the case with other chambered cairns, that the bones of the deceased might then have been stored within the tomb. Perhaps the eagles and ravens were thought to carry the soul or spirit of the dead to the other world, or to act as mythical protectors, interred along with the ancestors to guard them in the afterlife? The eagles may even have been a sacred totem animal symbolising the tribe and its ancestors. Other creatures may similarly have been adopted by neighbouring tribes, there being a site north of Isbister known as the 'Tomb of the Dogs'. Only eight human skulls were found here, alongside 24 dog skulls.

Isbister quickly came to be dubbed *The Tomb of the Eagles* (the title of Hedge's authoritative book) and now boasts a handsome visitor centre run by the Simison family. Ronnie and his wife Morgan no longer give guided tours but their daughters more than adequately carry the torch for them. I was already captivated by John Hedges' story when he contacted me for a photograph of a sea eagle to include in his book, but it was not until 2008 that I was able to marvel at the site for myself.

Sea eagle bones have been identified from many archaeological sites in Britain, some of them interglacial. In their encyclopaedic *History of British Birds* (2009), Derek Yalden and Umberto Albarella list 58 archaeological sites in England, with a few from Ireland and another cluster in Orkney, where Sea Eagle bones have been excavated. This confirms my own cursory survey in my 1983 book suggesting just how widely distributed the Sea Eagle used to be in prehistory.

Claws belonging to the species, along with many other waterbirds, have been found in Tornewton Cave in South Devon, from the third or Ipswichian Interglacial and the succeeding warm phase, some 130,000 years ago. The ice sheets in Britain advanced again during the Devensian glaciation, but during this time there were several brief mild periods, during one of which deposits were laid down at Walthamstow in Essex, for instance. Here the bones of greylag goose, mallard and tufted duck have been found, together with the left tibio-tarsus of a White-tailed Sea Eagle. In post-glacial times, remains of the eagle become more frequent, and of course there is no reason why more will be not be found in the future. For the moment we have a claw from the Gower Peninsula in south Wales and bones from Cheddar Gorge dating from some 1,300 years ago.

Both Mesolithic and Neolithic sites feature with Bronze Age remains from Salisbury Plain, the Fens and Glastonbury on the Somerset Levels, the latter two being wetland sites. Similarly, bones have been excavated from Co. Meath and the Iron Age lake village at Glastonbury. The ornithologist and historian James Fisher listed over 40 different species from Glastonbury – Cormorant, Heron, Dalmatian Pelican, ducks, geese, swans, Coot, Crane and gulls, all potential Sea Eagle prey and in ideal habitat, just as is found in central Europe today.

Other Iron Age and Roman sites in southeast England have turned up remains of Sea Eagles. A bone was also found in Dun Vulan, the Iron Age broch on the machair of South Uist near my home, while eagle bones, including a talon, have been excavated from Jarlshof in Shetland from the same period. And, of course, the Sea Eagle would seem to have been common in Orkney around this time.

Orkney's prehistory took on new significance for me back in 1975, after attending my younger brother's marriage to a local Stromness girl. I was able to visit Kirkwall's Tankerness

Museum where I saw displayed an 8th-century Pictish symbol stone from the Knowe of Burrian, now the museum's logo. I was already familiar with eagle symbols on several Pictish stones from the north of Scotland, but so crude were they that never had anyone attempted to assign a species to them. Most are highly stylised, or have been much modified over time. But the beautiful outline on the Burrian slab immediately jumped out at me as one of the finest depictions of a Sea Eagle I have ever seen. The carver obviously knew his subject intimately and had accurately captured the burly, vulturine aspect of the Sea Eagle, together with its thick, ragged plumes of wings and tail, and the massive head and beak.

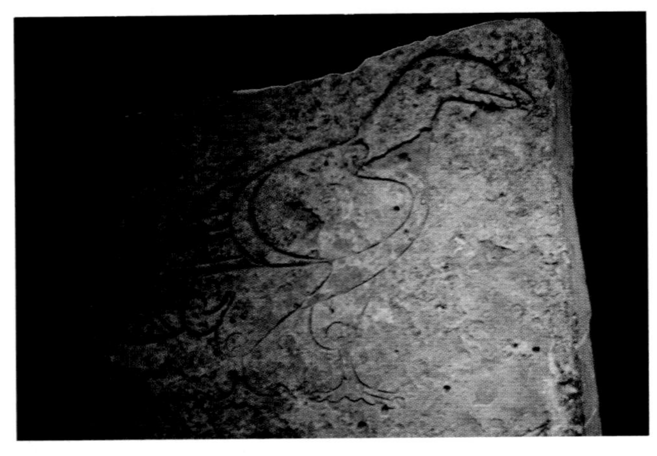

Pictish eagle carved in stone at the Knowe of Burrian, Orkney and now in Kirkwall Museum

But most characteristic of all were the bare tarsi. The Golden Eagle – the only other possible contender – has feathers all the way down to its toes. Roundng off the whole wonderful image were scroll-like adornments at the shoulder. Only recently, I encountered what seems to be the lower half of this same design on a broken stone in the Gairloch Museum in Wester Ross.

Although quite stylised, other Pictish eagles, when depicted with a fish, may be assumed to be Sea Eagles. A fine example in St Vigean's Museum, Angus, shows one in the act of devouring a salmon while another – originally from Latheron in Caithness and now in the National Museum of Scotland – stands over a fine salmon.

In Celtic mythology the eagle, occasionally perched on a sacred oak tree, is often linked to the salmon, the fish of knowledge. An eagle, often clutching a huge fish, was adopted in mediaeval gospels as the symbol of the evangelist St. John, one of the finest being in the supremely decorated manuscripts of the famous Book of Kells in Trinity College, Dublin. It is thought this meticulous devotional undertaking was begun in the Isle of Iona in the Hebrides, only to be taken to safety in Ireland when the Vikings arrived. Intriguingly, our Burrian eagle (surely a direct copy, or was it the other way round?) reappears as a coloured design in a few 8th-century decorated manuscripts such as MS 197 in the Corpus Christi collection, Cambridge and one of the Northumberland gospels.

Outlines of Sea Eagle (top), the Knowe of Burrian eagle and Golden Eagle (bottom)

The Anglo-Saxons occupied much of southern Britain around this time, and were familiar enough with the Sea

Eagle to make frequent reference to it in their poetry. We are told how, prior to the Battle of Maldon in Essex in AD 991, when the Anglo-Saxon army made a heroic but ill-fated stand against the Vikings:

> ...a cry went up.
> the ravens wheeled above,
> the fateful eagle keen for his carrion.

A similar stanza from the 7th-century poem Judith took it one stage further:

> ...the dewy-feathered eagle,
> hungry for food;
> dark-coated, horny-beaked
> it sang a song of war.

Yet another poem in the Anglo Saxon Chronicle described the massive and bloody Battle of Brunanburh, said to have been 'the greatest single battle in Anglo-Saxon history before Hastings'. It was probably fought somewhere in the Wirral, Cheshire in AD 937. The army of Aethelstan, the Saxon king of England, slew five kings and seven earls amongst the rival host of Strathclyde Britons, the Scots of King Constantine II and Viking Irish. The victors:

> ...left behind them
> the black-coated raven, horny-beaked
> to enjoy the carrion.
> and the grey-coated eagle, white-tailed,
> to have his will of the corpses.

Similar references are to be found in the contemporary Viking Sagas from Iceland. From the 10th-century Egil's Saga:

> ...the eagle scans
> The battlefield

while two centuries later, King Olaf Trygvesson's Saga told how:

> They who the eagle's feast provide
> in ranked line fought side by side.

Sea Eagle depicted in an 8th-century Northumbrian gospel

An old print of a Sea Eagle

During the Dark Ages the landscape of Britain had been little modified by man; huge tracts of deciduous forest flourished on the heavy lowland soils and extensive marshland existed around the Essex coast, the Fens, the Broads and the Somerset Levels. Such habits would have been as attractive to Sea Eagles as they still are in many parts of Europe – the Baltic coasts of Germany and Poland, the complex river systems of the Danube and the like – where populations of this species persist to this day. Waterfowl and fishes such as pike abound in these lakes and marshes which are flanked by tall oak or beech stands or conifer plantations to provide suitable eyrie sites.

Thus we can assume that the White-tailed Sea Eagle was once widespread, if not common, even, in Lowland Britain and Ireland. Neolithic, Iron Age and Anglo-Saxon farmers had already begun to clear the lowland forests and to drain the fenland, processes which accelerated as the human population increased. On the newly-created farmland, 'noyfull fowls and vermins' were not tolerated and soon laws were introduced to encourage their destruction.

In his definitive book *The Atlantic Islands* (1958), Kenneth Williamson – the first Director of the fledgling Bird Observatory on Fair Isle, who had spent World War II in the Faroe Islands – mentioned how the Faroese believed that the claws of the Sea Eagle could cure jaundice. He also described a very ancient tax system called 'Nevtollur'. The custom demanded that every man with part ownership in a boat was obliged to pay yearly either one eagle, raven or like bird of prey to an officer in the capital Torshavn. The eagle was especially valued, and the man who presented its beak was exempt for life from paying further 'bill tax'. Indeed, this bounty scheme may have played no small part in the early extinction of the Erne in the Faroes. The pursuit of White-tailed Sea Eagles nesting on the impressive cliffs of the Faroe Islands would have posed little problem for a people whose economy was based upon fishing and scaling the steep cliffs to harvest seabirds.

In 1981, inspired by Ken's account, I finally achieved a long-held dream of visiting Faroe for myself, even witnessing one of their controversial pilot whale hunts. Along with two friends, I camped overlooking Tindholmur, where the distraught mother had climbed into an eagle's nest, as recounted at the end of Chapter 3. The next morning we sailed past it on the way to the lonely island of Mykines:

> Sunday, 23 August 1981: ... The vertical seaward face of Tindholmur was breath-taking; surely it could not have been this that the woman climbed!? Fulmars were nesting on it . . . It was a bit rough over to Mykines but the sea was alive with gulls, fulmars and puffins. The top of Mykines was in cloud now and it had dulled over considerably. . .
>
> Monday, 24 August: ... the Mykinesholmur gannetry was small compared with those in Britain, the young having been harvested until recently. Mostly they seemed to nest on the broad soil-covered ledges between the two lava flows (which are here called 'hamrar') and thus one or two, perhaps three, levels only. Fulmars were everywhere and many Puffins, considering the lateness of the season. We met Mrs Joensen coming back from the Puffin colony with several dead Puffins in her hand. . . She said that 18 people live on the island now; there were 100 here during the war, and 150 after it. Three fatalities had occurred on the cliffs in recent years – two local men harvesting seabirds and a visitor from Westmanna near the cliff path. When a woman loses her husband she tends to give up island life altogether, we were told.

At least a few pairs of Sea Eagles bred in Faroe up to the middle of the 18th century when, rather surprisingly perhaps, given the abundance of seabirds, they became extinct. Names such as Arnafjord, Arnadalstindur (717 m) on Stremøy and Arnafjall (722 m) in northwest Vagar are all that remain to indicate their former haunts, and few vagrant Sea Eagles have been seen – the last, I think, was in 1902. Perhaps, one day, some will return and become established once more. The species is already making a slow recovery in Iceland: from only 20 pairs in the 1960s, there were 66 pairs established in 2011, 43 of which bred producing 29 young (Ævar Petersen, pers. comm.) and an equivalent number has now re-established in Scotland.

In 1957 Ken Williamson was despatched to the newly-designated National Nature Reserve of St. Kilda as its first warden. He provided an effective liaison between the Nature Conservancy, the National Trust for Scotland, which owned St. Kilda, and the Royal Artillery, who were establishing a radar station there as part of the Uist Rocket Range. As well as migrants, Ken studied the unique St. Kilda Wren and Field Mouse, not dissimilar to the subspecies he had known on Fair Isle, and the breeding Snipe which, with his museum experience, he recognised as being akin to the race on Faroe.

As long ago as 1697 Martin Martin mentioned a pair of Sea Eagles that nested at the north end of St. Kilda. In 1895, the nest was located on the 530-m cliff of Conachair, the highest sea cliff in Britain. The pair had ceased to breed sometime between 1829 and 1841 which, just as on Faroe, is an early date for such a remote but suitable territory. St. Kilda is one of the premier seabird stations in Britain, if not Europe, and again, just as on Faroe, the Sea Eagle's disappearance from St. Kilda was in all likelihood linked to the skill of the St. Kildan cragsmen, with whom it shared a diet of seabirds. The islanders probably saw them as competition (Love 2009).

The St. Kilda archipelago is much smaller in extent than Faroe and might have held only a single pair. Forty miles away from the closest other Sea Eagles, in the Outer Hebrides, such remoteness would have made it difficult for any to recolonize St. Kilda. Similarly, Faroe is several hundred miles from Iceland and the same distance from mainland Scotland; its tiny population of Sea Eagles would hardly have been viable and was too remote to attract fresh colonists. It is therefore not surprising that both island groups, with the islanders' seabird-based economies, should be amongst the first Sea Eagle territories in northwest Europe to disappear, undoubtedly due to human persecution.

Conachair, St Kilda – at 426 metres, the highest sea cliff in the UK

In a British context, it is only in the last three centuries or so that documentary evidence gives us a slightly clearer picture of the Sea Eagles' history, with a long-winded literary corpus of material from select Victorian naturalists and sportsmen. I ploughed through as much as I could find for *The Return of the Sea Eagle* and found it a surprisingly rewarding experience. Those grand old ladies of Scottish ornithology, Evelyn Baxter and Eleonora Rintoul, had been quite thorough in *The Birds of Scotland* (1953) and these two volumes made a wonderful starting point. They had freely dipped into Harvie-Brown's encyclopaedic *Vertebrate Faunas*, which he compiled with a selection of other authors, and Robert Gray's less reliable *The Birds of the West of Scotland* (1871). I spent many days in the Science Library of Aberdeen University and in the Scottish Ornithologists' Club library in Edinburgh. I freely accept that, inevitably, my efforts within Scotland proved more thorough than had my scrutiny of Welsh, English and Irish material. Full references for the following account can be found in my 1983 book.

I was conscious of compiling what I hoped would the first comprehensive and chronological narrative of the Sea Eagle in Britain and its ultimate demise. This was an essential prerequisite for the reintroduction project, albeit slightly after the event. We needed to know where Sea Eagles used to occur, how widespread and abundant they might have been and why they had become extinct. Furthermore, could we be confident that the deleterious factors no longer prevailed? I will now summarise the material from my 1983 book (now out of print) which has nonetheless stood the test of time to become a useful resource for later authors. I strive to incorporate some new material where I can and where it is of relevance.

The only record I know from Wales, a personal communication from Dr W. G. Hale, was that 'the last known Welsh pair' seemed to have nested at Carreg-y-Llayn, near Nevin on the Llyn Peninsula, around 1880. It is however likely that other pairs were nesting on the Welsh coast and more references may yet come to light.

A pair had nested on the high cliffs of the Isle of Man until the eyrie was destroyed in a snow storm in 1818 (Yarrell 1871). In England, Seebohm (1883) noted how a pair used to nest on Lundy in the Bristol Channel and that one was shot on passage on Lundy as recently as 1880, but it is likely that the species had ceased to breed there long before. It is also thought to have been Sea Eagles that nested on Dewerstone (Drewstaignton) Rock on the River Plym in Devon, probably sometime during the 18th century or even earlier (D'Urban and Matthew 1892). The ornithological literature sometimes confuses Ospreys and Sea Eagles but it is not unlikely that coastal eyries were mostly tenanted by the latter. An 'eaglet' was taken in 1780 from a pair nesting on Culver Cliff in the Isle of Wight (Yarrell 1871), which someone later came to dismiss as a Buzzard, but I see no particular reason to doubt that it was indeed a Sea Eagle.

More is recorded about Sea Eagles in the Lake District (Yarrell 1871; Macpherson 1892; Mitchell and Robson 1976). Unexpectedly, there is a mediaeval document, dated 1272, that decreed how the small tenants of northern Cumberland 'must preserve the nests of Sparrowhawks and Eagles'. Needless to say, this attitude did not persist. In the Lakes there is sometimes confusion with the Golden Eagle but it appears that at least one pair of Sea Eagles had an eyrie on Eagle Crag in Derwentwater occupied:

> ... from time immemorial . . . far removed from gunshot, and undisturbed by men; for no adventurous fool ever dared assail their lofty habitation.

Uncharacteristic and premature sentiments, for the later history of eagles in Lakeland is rather one of relentless persecution. From 1713 to 1765 more than 30 eagles were killed in Crosthwaite parish alone, with a bounty of sixpence being paid for a young bird and a shilling for an adult. Apparently the men of the parish kept a long stout rope at Borrowdale expressly for the purpose of climbing into the local eyries each year. At least one of these nests certainly belonged to a pair of White-tailed Sea Eagles; in about 1769, when a local farmer was lowered into the eyrie, the parent eagles circled overhead screaming anxiously (behaviour typical of this species, but not of the Golden Eagle). The nest was robbed again three or four years later, 'it being a common species of traffic in this country to supply the curious with young eagles'. Seldom a year passed without the Lakeland eagles being shot, or their eggs or young being taken.

Sea Eagle eyries are mentioned on Wallow Crag near Haweswater until 1787, Keswick, Ullswater and in Eskdale until about 1794. Two Sea Eagles were seen at Ullswater as late as July 1835 and this is often the date quoted in the literature for the last nesting by the species in England; there is, however, no record of this pair actually having attempted to breed. Golden Eagles ceased to breed in the Lakes at about the same time, although a stray was seen in 1860 or thereabouts. According to that redoubtable local ornithologist Ernest Blezard, this Golden Eagle met a tragic if honourable fate. A certain Farmer Jenkins rammed 'a terrible gurt charge' of powder and lead down his muzzle-loader and lay down on his back to take aim at the eagle as it flew overhead. He fired and, not surprisingly, suffered a broken collar bone from the violent recoil; the eagle was only wounded and 'gave further account of itself' before being finally despatched.

During the early part of the 19th century, however, the Sea Eagle was still reasonably plentiful in the remote parts of Ireland and Scotland, where the hand of man had little altered its coastal habitat. Here persecution was to become its sole concern.

Ussher and Warren's *The Birds of Ireland* (1900) was a useful starting point and I was able to trace at least 50 pairs of Sea Eagles once nesting in Ireland. Other pairs will have disappeared before records began and I know that my examination of Irish sources – and especially place names – was by no means complete. My figure was thus undoubtedly an underestimate. Indeed, Allan Mee (pers. comm.), who manages the reintroduction of Sea Eagles into Killarney, now reckons there were in excess of 75 nest sites in Ireland.

A few Irish pairs nested inland, with three or four pairs said to have bred, for instance, in the Mountains of Mourne (Co. Down) prior to 1831. Most, however, were coastal. Only a few are known from the east coast, for example at Saltee (Co. Wexford), Lambay Island (off Dublin) and Fair Head (Co. Antrim). But it was the west of Ireland that undoubtedly held most pairs. In Donegal eyries were known on Malin Head, Tory and Owey Islands, on Arranmore, Horn Head (two pairs) and at Teelin Head. In southwest Ireland a pair of Sea Eagles nested on the Blaskets in the 1870s, where on Inishvickallaun there is a place called the 'Hollow of the Eagles', with a 'Mount Eagle' on the adjacent mainland. Many other pairs must have frequented the cliffs of Kerry, and a further four pairs are known from the neighbouring Co. Cork – Sheep's Head, Bere Island, Crow Island and, prior to 1885, the hills of Berehaven. I only know of one site in Co. Clare, on the Cliffs of Moher, but undoubtedly there would have been others.

Off Erris Head is an 'Eagle Island', while no fewer than four sites, though not necessarily equating to four pairs, were known from Achill. Single pairs frequented the cliffs at Loughmuriga, Alt More and the Spinks, while three others nested on the great cliff of Alt Redmond. A pair was known to have bred on Inishboffin, with some 14 other pairs in Galway. But it was the huge precipices of Mayo which 'long afforded a home to this our largest bird of prey'. A recent review of evidence specifically for the County of Mayo (Lysaght 2004) traced about fifteen likely eyries so, assuming this situation to be typical of some other counties on the Atlantic coast of Ireland, the species must at one time have been pretty numerous there. It was on the coast of north Mayo that the last pair of Irish Sea Eagles are said to have bred, probably in the year 1898.

Within historic times it is clear, however, that Scotland provided the stronghold of the Sea Eagle in Britain. I have traced more than 100 pairs, meaning this to be taken as a minimum figure. I get slightly frustrated that this is often assumed to represent a total for Scotland. I perhaps should have stressed that more sites would come to light in the future, as indeed they have.

I have not included the full references here since they would overburden the text; many have already been mentioned and they can all be located in my book *The Return of the Sea Eagle*. In the 19th century, at least, the majority of Sea Eagles would seem to have been found nesting in the Hebrides, on the west coast mainland and in the northern isles of Orkney and Shetland. Occasional pairs were said to nest in the south of Scotland, most of them inland, but these lowland pairs ceased to breed at a comparatively early date, as did those in Lakeland to the south.

Eyries on two headlands in Wigtownshire ceased to be occupied by 1800 although Sea Eagles were still present in Galloway until 1836. Loch Skerrow is specifically mentioned as a territory 'until a railway [now disused] arrived to disturb its shores'. These Sea Eagles apparently then moved to Glen Trool where 'seemingly annoyed by the continued improvements... they retreated to solitudes more profound'. The ornithologist Duchess of Bedford claimed that the species last bred in the area in 1852, although Dick Roxburgh (pers. comm.) put it at about 1866 when a pair was trapped in the vicinity of Mullwharchar.

Sea Eagles used to breed on Ailsa Craig, where a straggler was shot in 1881, to be preserved in Culzean Castle. In 1812 the local newspaper reported how a pair on an island in a freshwater loch at Stair in Ayrshire had 'from time immemorial fixed his residence'. It is not surprising, after such publicity, that both eagles were trapped. Ernes were said still to be breeding in inland districts of Ayr and Dumbartonshire as late as 1840. Given the name, it is worth mentioning how the Eaglesham area possessed:

> ...several woods especially on the banks of the river [which] together with the rocks in the neighbourhood are much frequented by eagles.

Apparently they would often perch on the holm or low ground where the village was later built and it was further claimed, indeed, that it was they who gave Eaglesham or Eagleholm its name. We have already seen, as most authorities would agree, that this most likely derives from 'Eccle fia-holm' meaning 'the church in the hollow'. The lochs to the southwest of Eaglesham itself would certainly seem an ideal habitat for Sea Eagles, while nearby runs a small river called 'Earn Water'. Again etymologists would aver that this could derive from 'earn' meaning 'white', but these accounts do seem to amount to a remarkable coincidence. And there, sadly, the matter might have to lie.

There are a couple of 'Earn Craigs' in Nithsdale where Sea Eagles bred until 1837, and a possible site in Eskdalemuir. In his poem *Marmion*, Sir Walter Scott immortalised Loch Skene where 'eagles scream from shore to shore'. These would seem to have been Sea Eagles rather than Ospreys and may have been the same pair which frequented Talla Linnfoots in adjacent Peebleshire. Every effort was made to extirpate them but in 1834 they reappeared and 'committed several depredations' on sheep. An eagle was once found dead on Moffat Water with its talons firmly embedded in the back of a salmon, while a pair continued to nest on Loch Skeen until 1908.

Not quite in the same league as Scott, the poet Leyden suggested that Sea Eagles may once have bred in Roxburgh, on:

> Dark Ruberslaw . . .
> where perches, grave and lone, the hooded Erne…

It is interesting, too, that early in the 7th century St. Cuthbert is said by Bede to have encountered an eagle in Roxburgh. The bird had just caught a fish which the saint presumed to have been an offering for him from God, but typically the good cleric left the eagle half a share. This reminds me of a similar encounter I had with a Golden Eagle in Rum in the 1980s. After feeding the captive Sea Eagles, I drove down to the bay of Harris on the island's southwest shore and flushed a Golden Eagle from the raised beach nearby. I noticed something drop from its talons so, being interested in the diet of Golden Eagles where there was a lack of rabbits and hares, I wandered over to investigate. The prey turned out to be a red grouse, freshly killed and still warm. Never having tasted grouse before, and having nothing for my dinner that night, I decided to take it home. The eagle, however, had only flown a short distance away so I hurried back to the Land Rover for a gash haunch of venison intended for the Sea Eagles the next day. Perhaps having St. Cuthbert's gesture at the back of the mind, I left the meat on the knoll where the grouse had been dropped. Watching from a safe distance I was delighted to see the Golden Eagle quickly return to its perch and I swear I detected a look of astonishment on its face at the generous offering it found there.

Place names in Berwickshire and neighbouring counties hint at ancient inland haunts of the species – 'Earnscleugh hill' and 'Earnscleugh Water' in the Lammermuirs near Lauder, for instance, and 'Earnslaw', south of Duns. The cliff 'Earnsheugh' above the impressive seabird colonies of St. Abb's Head was almost certainly an old nest site. James Fisher (1966) was of the opinion that the 7th-century Anglo-Saxon poem *The Seafarer* referred to the Bass Rock:

> There heard I naught but seething sea,
> Ice-cold wave, awhile a song of swan.
> There came to charm me gannets' pother
> And whimbrels' trills for the laughter of men,
> Kittiwakes singing instead of mead.
> There storms beat upon the rocky cliffs;
> There the tern with icy-feathers answered them;
> Full oft the dewy-feathered eagle screamed around.

In the distant past, gannets bred at Trevose Head in Cornwall and on Lundy, and these might constitute the only other likely contenders. Both Bempton Cliff in Yorkshire (c. 4,000 pairs) and Troup Head on the Moray Coast (more than 1,500 pairs) are only recently established (1937 and 1988 respectively). But the Bass remains the only significant Gannet colony known in southern Britain. John Wolley casually mentioned how this famous volcanic plug was indeed once a haunt of the Erne so it would seem that Fisher could be right. It is therefore tempting to reflect that the Sea Eagle territory on the Bass could have been tenanted for a thousand years, or indeed may even pre-date the Anglo-Saxons.

Bass Rock, East Lothian

Further north still, in Kincardineshire, there lies another cliff called 'Earnsheugh', just to the south of the city of Aberdeen. The name 'Earnskillies' in Glen Clova may indicate an inland eyrie which seems to have been occupied in 1813. Prof. William Macgillivray, an outstanding field naturalist heading the Natural History Department at Aberdeen University, believed that a pair of Sea Eagles once bred near Lochnagar (a mountain, despite its name!). Several other potential sites have now come to light. Formerly, eyries were known to have been located on Troup Head and on Pennan Head on the south coast of the Moray Firth, both notable seabird colonies today.

Pennan Head is associated with an ancient prophecy of Thomas the Rhymer who, in the 13th century, is reputed to have declared:

> There should be an Eagle in the crags while there was a Baird in Auchmedden.

When the Baird family finally sold their Auchmedden estates some time later, the eagles ceased to breed on Pennan Head. They returned, however, as soon as the new owner married Miss Christina Baird. The estate later passed out of the family once more and the eagles deserted the site once and for all. This tale seems to suggest one method by which Sea Eagles could be encouraged to breed on the Moray Firth once more, if only a suitable marriage could be arranged. However, in 1903, G. Sim (1903), an Aberdeen tailor, naturalist and taxidermist, implied that this legend is more appropriate to Peregrine Falcons. On the other hand, the late Dr Ian Pennie told me of a similar story relating to the Ernes nesting on Dunnet Head in Caithness.

The notorious collector and sportsman Charles St. John (1849) located an eyrie on the north side of the Moray Firth, at the Soutars of Cromarty in Easter Ross. None of these North Sea coastal sites could still have been occupied by the early 19th century and, in view of the very tentative nature of the evidence for the existence of some of them, they may be very much older than that.

We will now resume our survey back on the west coast where, in 1849 a young Sea Eagle was taken into captivity from a nest in Catachol Glen on the Isle of Arran. Robert Gray added:

> Sometime during the ensuing winter the ledge of rock on which the eyrie was placed gave way, and the eagles – probably viewing the mishap as a timeous warning – left the district.

Apparently a pair did build a nest in the same glen in 1870, the female sitting on two eggs before deserting.

Rannoch and Blackmount, Argyll

On the Argyll mainland, just to the north, the flooded moors of Rannoch and Blackmount, rich in freshwater lochs, had much to attract Sea Eagles. In the 1830s a dozen or more might be seen sitting around on the stumps of old trees, and a local forester Peter Robertson – a 'model Highland deer-stalker' from Blackmount in Argyll, who died in 1877 – recalled 'packs of over twenty birds'. Doubtless such gatherings, so typical of this social species, included many juveniles or overwintering birds, as only five nests are known in the area. One or two of these may have been alternative sites of the same pair. All of these nests were constructed from sticks in small trees on tiny islets on the lochs, except for the cliff eyrie above Loch Atriochan, possibly the site where two eggs were collected in 1866.

Although an egg collector and sportsman typical of his time, John Colquhoun (1805–1885) eloquently described one particular expedition, when he spotted a female Sea Eagle at her eyrie in Rannoch:

> . . . her white tail shining like a silver moon. . . When we neared the islet, they both flew out to meet us, uttering their shrill screams. Sometimes they floated at immense height, and then, cleaving the air in their descent, flew round the eyrie beating their wings which made a hoarse, growling noise . . . [Their nest] was very large sticks and twigs.

Peter Robertson told Colquhoun of a shepherd lad who had robbed the nest in May 1850:

> He has swam in at nicht, the scoondrel, and ta'en the eggs or young for fear o' his lambs... Many a time he has swam [across] Loch Rannoch in the night-time to see his lass.

Robbing the Eagle's Nest by R. R. McIan

Sea Eagles were also said to frequent the coastal hills of Creran and Etive and on the Mull of Kintyre, but none of these would appear to have survived beyond about 1870. Ardnamurchan Point was deserted around 1890, although a pair of adults there was spotted from the mail steamer as late as April 1913 and could well have been breeding.

Moving north on the Inverness-shire mainland, an isolated pair was known near Loch Laggan, although there are also records of Ospreys breeding there. It is not inconceivable that the two species might use the same nest, although obviously at different times. Another pair of Sea Eagles nested on one of the islands in Loch Loyne until the fir tree was felled in 1835; thereafter the eagles nested in a birch. Here also there may have been some confusion with Ospreys which were said to have bred on the loch as late as 1916. The bird artist and author Mrs Jemima Blackburn mentioned that Sea Eagles were 'not very uncommon' around Inverailort in West Inverness-shire, where she sketched one eyrie on 20 June 1864; the single chick had its beak deformed by a trap placed in the nest for the adults. Although her cliff looks a bit too spectacular for this part of West Inverness, this same site may have been the inspiration for another artist, R. R. McIan, and his melodramatic colour plate entitled *Robbing the Eagle's Nest*. Note not only the obligatory lamb in the nest but also two large cod beside the twin chicks.

I have visited scores of Sea Eagle eyries in my time and have never been attacked as McIan depicts. William Macgillivray agreed:

> . . . a man may carry off its young before its face with little danger of a scratch.

He did mention, however, a man in Lewis who was struck with a wing from behind as he lay to shoot eagles at the nest. Similarly, Harald Misund, who has visited countless more eyries than I have, wrote to tell me how he was only once 'attacked' at a nest in Landegodë near Bodø. The bird came from behind and swooped quite close to his head. He had visited that nest many times but had experienced nothing like that before.

There is little precise information available for the northwest mainland of Scotland and in Wester Ross, Sea Eagles do seem to have been relatively scarce. Around Gairloch, Osgood Mackenzie mentioned a keeper who shot three adult eagles on Baos Bhheinn, together with two young in the eyrie – probably making them Sea Eagles – and all before breakfast. There was also a well-known eyrie at Fionn Loch atop a crag on a small island known as 'Eilean na h-Iolaire'. As soon as a boat was put on the loch for anglers, the eagles were forced to move to Beinn Airidh Charr in the Fisherfield Forest, a nest which ceased to be occupied in about 1850. Another eyrie was located on a sea cliff in Loch Broom opposite Ullapool.

The Sea Eagle was more common in Sutherland and 'abounded' around Kildonan, Clyne and Edrachillis. Charles St. John claimed that Ospreys bred on Loch Meadie but Wolley disputed this, considering the eyrie construction and some feathers he found nearby to be those of the Sea Eagle. They both could have been right, of course, depending upon which species had claimed the nest that season. Ospreys, though, would probably defer to the larger Sea Eagle, according to experience in Sweden (Björn Helander, pers. comm.)

It was probably while stealing eggs somewhere in Sutherland that John Wolley paused to pen a particularly eloquent description of that Sea Eagle's nest:

> To enjoy the beauties of a wild coast to perfection let me recommend to any man to seat himself in an Eagle's nest. [It was] on a sort of triangular ledge, a small Rowan tree touching it in front. The rock is scarcely overhanging. The nest is made chiefly of dead heather stalks, with a few sticks for the foundation, the largest of which are above an inch in diameter and two feet long. It is lined with a considerable depth of moss, fern, grass and Lazula . . . The hollow is small for the size of bird, and very well defined. There is a rank sort of smell, but no animal's remains in or near it; several feet below me is an old nest.

The famous seabird cliffs on the island of Handa, now a Nature Reserve, were said to have been vacated by Sea Eagles in about 1864 but John Colquhoun found eaglets in the eyrie in 1877.

In 1871 Robert Gray summarised:

> On the mainland the breeding localities are much less numerous than in the islands; there are still, however, although I have no wish to see their privacy invaded, a number of frequented eyries in several of the counties stretching from Cape Wrath to the Mull of Galloway.

In 1814 Sir Walter Scott accompanied the famous lighthouse engineer Robert Stevenson on his annual inspection cruise. Passing Cape Wrath, the northwest tip of mainland Scotland, he observed:

> This dread Cape, so fatal to mariners, is a high promontory whose steep sides go down to the breakers which lash its feet...I saw a pair of large [sea] eagles, and if I had had the rifle-gun might have had a shot, for the birds when I first saw them, were perched on a rock within about sixty or seventy yards. They are, I suppose, little disturbed here for they showed no great alarm.

In 1879 John Colquhoun shot an adult at one of two known eyries on the high cliffs between Clo Mor and Garbh Island to the east along the north coast of Sutherland. According to his son, this was the only Sea Eagle that Colquhoun ever shot. Sadly, though, its widow never found another mate. In May 1852 John Wolley obtained two separate clutches on Whiten Head nearby, both his eyries being only 1.5 miles apart.

In the relatively low-lying county of Caithness the Sea Eagle was said to have been more common than the Golden Eagle, but I can trace only three eyries. The eyrie on Dunnet Head was popular with egg collectors. It was said that in 1840 an eyrie had been placed on the larger of the two stacks of Duncansby, and some time before 1887 a pair bred at the great seabird colonies on the Ord near Berriedale.

Turning now to the west coast islands, the apparent scarcity in the records of Sea Eagles amongst the Southern Hebrides of Argyll may have been a consequence of a lack of sizeable seabird colonies. Several Sea Eagle territories were to be found on the inaccessible precipices of Jura's west coast while on Islay, next door, nesting may have occurred into the early years of the 20th century. Mull, of course, is now a popular spot for the reintroduced Sea Eagles and, not surprisingly, once held several nesting pairs.

Stacks of Duncansby, Caithness

To the north, three of the four Small Isles were well endowed with Sea Eagles: a pair in Canna until 1875 and two or three pairs on Eigg until 1886. The mountainous Isle of Rum, the largest, had five pairs of 'eagles' but I suspect this included three pairs of Sea Eagles on the coast and maybe two pairs of Goldens inland. By the start of the 20th century, Sea Eagles were said to be 'very rare' in Rum, although a single pair persisted until 1907 when two eggs were collected and one of the adults was shot. Although very few Sea Eagles were left in the whole country, the widowed bird somehow found a new mate until 1909, when both were shot at the same locality.

In the southeast corner of Rum is a rock wall called 'Sron na h-Iolaire' – the nose of the eagle – which, being close to the island's best seabird cliffs, would have made an attractive site for Sea Eagles. Between the two world wars, when the Sea Eagles had already been wiped out, this was used as a roost by Golden Eagles. Indeed, in recent years, two of the gamekeeper's gin traps have been found on ledges there, one found by Peter Corkhill and me in 1975, rusted solid in a set position. There is another eyrie on the northwest cliffs of Rum, near Wreck Bay. When George MacNaughton came to be gamekeeper in the 1930s, the site was occupied by Golden Eagles; he fixed a length of fence wire to the rock so that he could swing himself on to the rock ledge. Much later, Morton Boyd's son Ian repeated the feat to check the nest contents for me. Much to George's astonishment, Ian opted not to use the wire since it looked so rusty. In the end, George moved on in 1977 to become a dedicated Chief Warden for the Nature Conservancy's new Rum National Nature Reserve, going one better than 'poacher turned gamekeeper'!

The Isle of Skye, just to the north of Rum, was according to Robert Gray:

> ... the headquarters of this conspicuous eagle in the west of Scotland – the entire coastline of that magnificent country offering many attractions to a bird of its habits. Nearly all the

> bold headlands of Skye are frequented by at least one pair of Sea Eagles… One of the most picturesque eyries of the Erne on the west coast is perhaps that placed on the breast of one of 'Macleod's Maidens', a group of three sharp-pointed stacks of rock on the coast of Skye.

Harvie-Brown and his co-author, the Rev. Macpherson, who lived in Skye, knew of seven eyries between Waternish and Loch Bracadale. A further nine or ten eyries were known to Gray from Gob na h-Oa, south of Loch Bracadale, to Loch Brittle, below the Cuillin. It is possible that some of these 17 nests were alternative sites only, but it is indisputable that the Sea Eagle was abundant on the west side of the island. In 1879 an eaglet was shot on its nest – on a basalt pillar on the bold face of Dunvegan Head – by a Mr H. Parsons Jr from Oxford. It was at this eyrie, too, that Captain Macdonald obtained the two fledglings that would become his 'familiar companions' described in Chapter 3.

A few pairs were also to be found on the east side of Skye, for instance on the Quirang. Here John Colquhoun asked the local keeper to obtain for him the complete eagle eyrie to add to his personal museum. The old Skyeman knew of ten pairs along that particular stretch of coastline, but this does seem a bit of an exaggeration. Doubtless, though, Sea Eagles nested on Raasay, just offshore, where there is a hill named 'Beinn na h-Iolaire'.

Baxter and Rintoul (1953) had a record of what is apparently, and is still accepted as, the last British breeding pair, which had a nest on the west coast of Skye in 1916. When they visited in 1930, however, they were also told of a pair that had bred long after 1916, although no evidence has ever emerged to support this claim. (McMillan 2005).

Today Skye vies with Mull as the main stronghold of the reintroduced Sea Eagles.

Around the year 1700, Martin Martin mentioned a Sea Eagle eyrie on the huge basalt columns of the Shiant Isles in the Outer Hebrides, and this was still tenanted nearly two centuries later when a yachtsman, anxious to show off his prowess with a rifle, shot both adult eagles (Dr Hugh Blair, pers. comm.). A mounted specimen in the possession of a Stornoway lawyer may well be this very bird.

Garbh Eilean, Shiant Isles

Several other pairs were known on the Isle of Lewis, the Park area being a local stronghold, with as many as 13 in a single year having been destroyed there in the 19th century by one local man. Towards the end of the 19th century, Edward Booth collected several clutches there. Booth was 'well supplied with means which he used lavishly in following his favourite pursuit of collecting birds'. He was more interested in skins than in eggs and recorded his experiences in his *Rough Notes*, published between 1881 and 1887. Ruthless in his quest, he was not averse to bribery and corruption to procure specimens. So obsessional was he that he often used false locations in his book to outwit his rivals. When he died at the age of 50, his collection passed to Brighton Museum, which also holds his diaries, a more reliable source of information.

The striking basalt columns of Garbh Eilean, Shiant Isles, where one of the last Sea Eagles bred

Again around 1700, Martin Martin mentioned both eagles in Harris where, 150 years later, Robert Gray knew of several eyries, one on the offshore Isle of Scalpay. Martin also knew of Sea Eagles in North Uist where eyries were later documented on the Lees, and on a low cliff near Baigh Chaise, deserted by 1880, on the island of Wiay with another on Benbecula. On South Uist a pair of Ernes nested where the lighthouse of Ushinish came to be built in 1857. C.M. MacVean had a pet Sea Eagle called Roneval, named after another hill in South Uist from where the bird had been taken.

The species was said to have bred on the islands to the south of Barra until 1869. Sea Eagles bred on a stack on Mingulay, appropriately named Arnamul, until about 1848. The people on Mingulay, who evacuated the island in 1912, were renowned fowlers and would even scale the cliffs for seabirds without the security of a rope. As a result, this spectacular island is often referred to as 'the other St. Kilda', so it is not surprising that Mingulay achieved early extinction of its Sea Eagles, just as St. Kilda and Faroe did.

Orkney and Shetland are other island groups where the people once harvested seabirds and, being a particular hotspot for Sea Eagles, they had a long association with the species, going back to prehistoric times. In 1693 Wallace wrote how in Orkney 'eagles or earns are here in plenty'. In the 17th and 18th centuries, eyries were known near Houton Head, Mull Head, Costa Head, on Rousay, possibly Eday, South Ronaldsay and Switha, where there are places called 'Erne's Knoll', 'Ernie Tooer' or 'Erne Tower'. But such sites on the low cliffs of Orkney, many of which were once regularly scaled to harvest seabirds, must have been exceedingly vulnerable to human disturbance and once again would explain their having been deserted at a comparatively early date. The only sites to survive into the 19th century would have been on the high sandstone precipices of Hoy, the Norse name meaning, appropriately enough, 'the high island'. As many as 12 nest sites were claimed but the actual number of breeding pairs must have been lower than this. They included the famous Old Man of Hoy, White Breast and inland near the Dwarfie Stone. Robert Gray described two young being taken from a nest on the west cliffs of Hoy in June 1812, probably close to the Old Man. He also quoted Bullock's vivid description of the:

> ...towering rocks rising to a perpendicular height of 1200 feet from the sea. About one third of the way up this awful abyss a slender-pointed rock projected from the cliff, like the pinnacle of a Gothic building. On the extremity is a hollow, scarcely of sufficient size for the purpose [which] these birds had fixed on as a place of security for rearing young; the situation was such as almost to defy the power of man to molest their habitation.

But, alas, not quite. In the latter half of the 19th century, clutches of Sea Eagle eggs were repeatedly stolen from the eyrie, as well as two young birds. A surprising conclusion to the saga of Sea Eagles in Orkney, according to the Jourdain Society, is that an egg was apparently collected there as recently as 1911.

The cliffs of Hoy, Orkney and the Old Man (far right)

In the past, Shetland rivalled Skye or the Outer Hebrides for the densest population of Sea Eagles in Scotland. In the definitive *Birds of Shetland* (2004), 20 sites were documented and while some of these may have been alternative nests, others may have been overlooked. The Rev. George Low from Orkney spent over two months in Shetland in the summer of 1774, and found all the islands' eyries were on precipices, sea cliffs or stacks, except for one built on a low rocky island in a loch at Northmavine on the Shetland mainland (Bobby Tulloch, pers. comm.). Locals would swim out to the holm on floats to destroy the eaglets.

Sea Eagles ceased to breed on Fair Isle between 1825 and 1840, the eyrie having been on Sheep Rock. The cliff names 'Erne's Brae' may simply have been an alternative of this same pair. On the equally remote Isle of Foula, an eyrie was built on the spectacular face of the Sneug, a cliff that is a close second to Conachair in St. Kilda as the highest in Britain. The islanders once protected the great skua because they claimed the Bonxie was the only bird bold enough to drive Ravens and eagles from their sheep. With over 2,000 pairs of Bonxies now breeding on Foula, I am sure the Foula folk would gladly settle for a pair of eagles again. It might have been Foula where two young were taken from an eyrie, although the attempt to shoot the adults failed...

> The parent birds never again occupied the same nest. In after years they tried one new place after another, but misfortune followed them until, from old age or some cause, they gave up breeding . . . They are now bleaching with age, childless and sad.

But, nonetheless, clutches of Sea Eagle eggs were said to have been taken on Foula as recently as 1901 and 1902.

In 1845 a Free Church minister visited Fitful Head, the southernmost tip of Shetland, and romantically described a pair of eagles 'soaring through the azure depths of the air'. The older local men admitted that the Fitful eagles never touched lambs. Nonetheless, with traditional but illogical efficiency, they repeatedly shot one of the pair. Its widow never failed to gain a new mate until ultimately both were killed that same winter. Thereafter no more bred on Fitful Head, although they are remembered locally by a rock in the Bay of Quendale nearby called 'Erne's Ward' – the eagle's lookout.

But there is an intriguing postscript to the story associated with this spectacular spot. A story appeared in the *Shetland Times* on 9 March 1984:

> James W Duncan... saw a pair of [of white-tailed eagles] nesting in June 1921... One of the eagles was in the nest while the other sat a short distance away. It was impossible to see if the nest contained any young birds.

Soon afterwards, Pete Ellis of the RSPB managed to track down the retired fisherman in Scalloway and was convinced that this was an accurate sighting. He wrote to Roy Dennis:

> Mr Duncan will be 81 next September. At the time of the sighting in mid-June 1921, the skipper of the *Sylvanus* LK171 went close under the cliffs... [and} spotted a huge nest made of dry tangles under an overhang about a quarter of the way down the cliff. This was on the north cliff near the Nev. One of the eagles was sitting on the nest so that only its head and shoulders were visible, and the other eagle was perched on a rock slightly above the overhang. The boat returned to the area at the end of August, the nest was still visible but no birds were seen. The size of both the birds and the nest obviously made a big impression on Mr Duncan and it is still a very vivid memory.

It is pertinent to mention here a letter that appeared in the *Daily Mail* on 23 April of, I believe, 1918. The correspondent –- 'a distinguished observer' – was none other than Lt. Commander J. G. Millais (1865–1931), the noted sportsman, horticulturalist, travel writer, naturalist and artist. He gained his rank in World War I when he served, as Vice-Consul in Hammerfest, with the Secret Service in Norway until he managed to escape in 1917. His letter referred to the report of the 'melancholy' disappearance of the last Sea Eagle in Shetland, which he hoped might yet reappear, just as it had in several seasons previously. But he went on to offer the following interesting observations, worth quoting at length:

> I doubt if it was the last of the residents, because until 1913 another single bird frequented the cliffs of Foula, and a pair nested regularly on the great precipice facing the Atlantic on the extreme north-east of Northmavine. This pair was protected by the late Mr Haldane of Lochend, and their nesting site, which I have seen several times, was deemed quite inaccessible until a bold egg collector assailed it about the year 1912.
>
> From what I have heard recently the pair of sea eagles frequenting the Hermadale cliffs are still there and should be strictly protected. I know of one other pair in the British Isles which have made their home on an island in the Outer Hebrides [probably the Shiant

> Isles] The place they have chosen as their home is quite inaccessible, and the nest has never been assailed, but there is always the danger of the birds wandering to the neighbouring islands and being shot when feeding on dead sheep in the winter.
>
> The sea eagle is now growing very scarce in Northern Europe…

Clearly, as in Skye, the status of the last Sea Eagles in Shetland throws up tantalising reports which may never be fully clarified. Eyries were known on Noss and the Bards of Bressay – undoubtedly alternatives of the same pair – while 'Erne's Hill' in the north of Bressay may have belonged to another pair. John Wolley had at least three clutches taken from here in 1847, 1856 and 1861. Ernes also nested on Papa Stour, each year 'from time immemorial' and in the north of Fetlar (probably the site of the heroic eagle-and-baby story recounted earlier) until 1895..

It seems that there were at least two resident pairs on Unst until about 1870, though Saxby stated that they ceased to breed there prior to 1859. One bred at Hermaness – the most northerly pair in Britain – and the other at Lund, where there is a rock called 'Erne's Hamar'. A fowler named Joseph Mathewson annually robbed the nest at Hermaness; on one occasion he caught the female sitting on the nest ledge and she was apparently stuffed for Buness House on Unst.

Several pairs may have frequented the island of Yell, for at the North Neaps there is an 'Eagle Stack' with an 'Erne Stack' on the east side near Aywick. The only documented site on Yell however – at the Eigg on the West Neaps, sometimes called the Neaps of Graveland

The Sea Eagle photographed by H. B. Macpherson in Shetland.
(Courtesy of the late Hon. D. Weir and the Macpherson family, Balavil)

– was the very last occupied nest in Shetland. The site had been known to Low as long ago as 1774 and doubtless remained in almost continuous use until 1904. In that year a single egg was stolen from it, eventually finding its way into the National Museum of Scotland. It is recorded that the robber was caught and punished but in 1910 the eyrie was again raided by a collector. Pennington *et al.* (2004) reproduce an eyewitness account:

> The clergyman… who robbed the Erne's nest on the Eigg on the west side of Yell… was called Sorby and he came from Derby. He got one addled egg in the nest, but was subsequently 'given away' by someone who assisted him in the climb to the eyrie, and fined £5 in the Sheriff Court in Lerwick and ordered to forfeit the egg. It was not until two years later that it was discovered that he had submitted a Golden Eagle's egg instead, and had retained the egg of the White-tailed Eagle!

His climbing spikes were long to be seen on the cliff face. In a 1904 issue of *Scottish Naturalist*, J. Tulloch gave some background to the effect that in 1903 the female was shot feeding on a dead sheep and only later died of her wounds, to end up stuffed by a local taxidermist. The male then disappeared but reappeared 14 days later with a female, but Tulloch then makes a comment that puzzles me: she was 'much smaller than itself and an 'albino''. We all know that it is the male that is the smaller sex, so are the sexes of this famous last pair – as we have always accepted them – really the wrong way around? Was it the male who was the 'albino'? Were these early observers too casual in allocating the sex of these birds? Tulloch places 'albino' in quotes and indeed the bird was not an albino. George Lodge, who also saw the bird, noted that its primaries stood out as a light brown colour. Pale, leucistic plumages in birds are not unknown, even among Sea Eagles (Björn Helander, pers. comm.)

Close-up of the Sea Eagle by H.B.Macpherson in Shetland. (Courtesy of the late Hon. D Weir and the Macpherson family, Balavil)

Whatever its sex, the alleged 'albino' was probably the same bird which had latterly frequented North Roe across Yell Sound. These sites may have been alternatives for the pair, and apparently the North Roe pair had two or three different nest sites while two of the five pairs still remaining in Shetland towards the end of the 19th century 'not only change from one district to another but even from island to island'. Thus the confusing reports of an 'albino' occurring both on Yell and North Roe might be explained. In 1910 'she' was widowed but returned to 'her' nest on North Roe 'to gaze out over the wide horizon and wait'. No new mate arrived, however, for in 1916, far to the south in Skye, the last known breeding attempt had already been made.

Tulloch noted that:

> Every inducement is offered in the way of money for the 'albino' dead, but a great interest in them is taken by the people of the vicinity who will not destroy the birds.

Nonetheless, in 1918 an old man is said to have shot this albino 'female', who seems to have been known and protected locally for nigh on 30 years. Lodge does not say where he learnt this last bit of information about the age of the bird; it is not implausible as, perhaps, it was probably known from another location but, according to Tulloch, it only turned up at the West Neaps in 1903 and died in 1918, fifteen years later.

Sometime between 1912 and 1918, Harry Brewster Macpherson from Balavil near Newtonmore in Strathspey made a special journey to photograph it. A few years earlier he had published *The Home Life of the Golden Eagle*, a collection of pioneering photographs taken from a hide at an eyrie in the Cairngorms. But then he took what is undoubtedly the only photograph ever taken of a truly British Sea Eagle, the last of its race indeed. I received a print from the late Doug Weir. Incidentally, since I wrote an article about it in Scottish Bird News in 2008, I have re-examined the accounts of the bird and the photo and now think that the bird does indeed look more like a male than a female. But, no matter: this bird's sad death marked the final extirpation of the Sea Eagle in Britain.

Chapter 6
Sea Eagle persecution

It is impossible to conceal the fact that if the present destruction of eagles continues we shall soon have to reckon this species amongst the extinct families of our 'feathered nobility'.
Robert Gray 1871

Over the centuries, the White-tailed Sea Eagle has been the subject of human persecution but, in southern Britain, this is likely to have taken second place to habitat loss, mopping up the last remaining pairs in England and, presumably, in Wales. The last nesting on the Isle of Man was in 1818. A single isolated pair on St. Kilda and another on Fair Isle disappeared around 1830, doubtless suffering like the Faroese pairs at the hands of local cragsmen and fowlers. In the north, many of the Orkney eyries must have been in ridiculously accessible locations and, in common with Galloway, would appear to have become deserted around the same time. Harvie-Brown and Buckley (1891) almost apologised for being:

> ...unnecessarily full in our notes on the Orkney eagles, but as a breeding species they are now quite extinct and rarely occur even as a passing migrant. Indeed, the Sea Eagle is rapidly disappearing all through Scotland, so it behoves naturalists to try and make their memorials accurated and full, seeing that, in the Orkneys at least, this is all that is left to us.

It was not until the 1840s that a more serious and rapid decline began. Although the celebrated pair at Loch Skeen in Dumfries-shire had gone by 1837, one or two others continued to breed in southern Scotland until about 1866. By this time, breeding had ceased on Arran, the last recorded date having been 1849, at the Soutars of Cromarty (1845), Portree on Skye and Lochs Fiag and Fionn in the northwest Highlands. These were all nests close to human habitation or else too easily reached by man. But by the 1860s more remote sites were becoming vulnerable, in such scattered localities as Unst in Shetland, the Lees in North Uist, Barra and Glencoe. During the next decade another North Uist site (Bagh Chaise), two in Rannoch (Lochs Ba and Laidon), two others in the Small Isles (Canna and Eigg), one on Mull (Burg) and another on Garbh Island in Sutherland could all be removed from the list. The 1880s saw the abandonment of two or three sites in Assynt, another on Harris and at least two pairs on Skye.

The last Irish pair would seem to have bred in 1898, by which time breeding had also ceased on both Fetlar and Bressay in Shetland.

But two or three pairs survived in Shetland into the 20th century (Yell and North Roe were possibly the same pair with another isolated in Foula). So, too, did two others in the Outer Hebrides (Ness and the Shiants) and one pair on Rum. A pair had been seen at Ardnamurchan as late as 1913, although breeding was said to have ceased there by 1890. There are records of a clutch taken in Sutherland in 1901 and others from Foula in 1901–1902. Sea Eagles were still breeding in Rum until 1909, while yet another egg seems to have been taken on Hoy, Orkney in 1911. The last reported nesting is said to have taken place on Skye in 1916 although a pair might have nested near Fitful Head in 1921 (Ellis, pers. comm. and Chapter 5).

It is obvious, too, from the previous chapter that many sites – 45 to be precise – are attributed fairly particular dates for their final abandonment. Assuming those dates to be correct, although it is likely that some are not, I have produced a chart of the decline of the Sea Eagle in Britain. I have also assumed that each site, or its alternatives nearby, was occupied continuously previous to its demise. This is not unreasonable since the Sea Eagle is known to be highly traditional in its use of nest sites, and indeed several – such as at Straiton in Ayrshire, Papa Stour, Yell and the Shiant Islands for example – are expressly described as having been occupied 'since time immemorial'.

As each eyrie is said to have been deserted it is removed from our total of 45. Some dates may not have been sufficiently precise so time periods of ten years are used. There are, of course, few dates available prior to 1800 but I would propose that up to this time any decrease in White-tailed Sea Eagles was a slow and gradual one, taking place over many centuries. Our curve, if it were taken to be representative, shows that the rate of disappearance of Sea

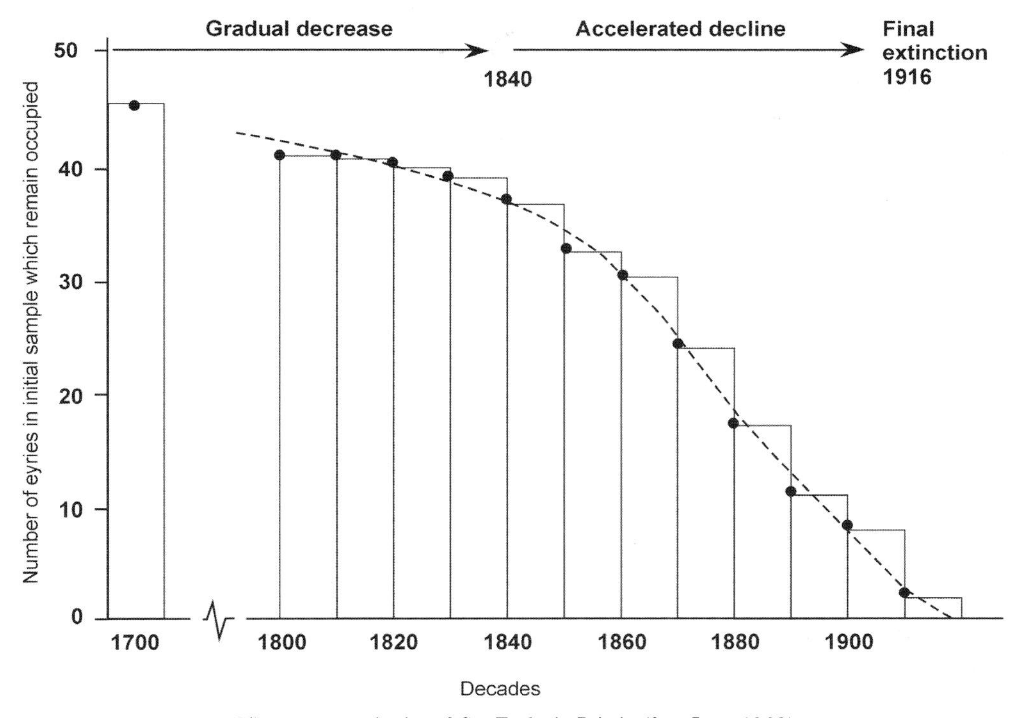

The apparent extinction of Sea Eagles in Britain (from Love, 1983)

Eagles in north and west Britain only gathered momentum from the 1840s until, by 1900, only a handful of pairs were known, all of them in Scotland. Within two more decades the species had become extinct.

What might have brought about the accelerated decline of the Sea Eagle? The habitat along the western seaboard of Britain would appear to have deteriorated over several hundred years but then changed little during the 19th century. There would have been many fewer trees than now exist in northern Norway, for example, reducing the number of suitable nest sites. But the main deforestation in the Highlands and Islands would have occurred long before the Sea Eagle went into terminal decline. Furthermore, deciduous trees contribute considerable enrichment to freshwater systems and it is likely that fish abundance began to decline as a result. Such an impact would have been felt most by the Osprey, and only the few inland pairs of the larger Sea Eagle. Also, coastal fisheries would probably have declined due to over-exploitation. For a variety of other reasons, though, the north and west of Scotland still offered the best hope for Sea Eagles, but only without the intense persecution that was to develop.

It is known that in a period from about 1550 to 1800 Europe had experienced a marked deterioration in climate; geographers often refer to this as 'the Little Ice Age'. Frequent cold winters and wet summers persisted until the 1890s when temperatures at last began to ameliorate. It is unlikely that the White-tailed Sea Eagle could have suffered directly from this climatic change since the species has such a vast geographical range, spanning such a diversity of latitude. Indirectly however, it could have been affected through changes in food supply. The spectrum of prey taken by the White-tailed Sea Eagle would seem far too wide for it to be adversely affected by the loss of a few, albeit much favoured fish species. It has been suggested that warmer ocean currents might have brought about an increase in other fishy resources on the prey list of Sea Eagles, in turn benefitting seabirds such as Gannets, Fulmars and Shags, among others.

That said, only the Gannet is currently still increasing. In recent years, auks, Kittiwakes, Shags and other seabirds in Britain have suffered almost catastrophic declines. Sand eels, their main prey, seem to have been impacted by climate change. As our seas have warmed, the eels have retreated northwards while other species, notably pipefish, have moved in. Seabirds are then forced to bring these unsuitable fish back to their young. Pipefish, being long and thin with spiny bodies and lacking the oily flesh of sandeels, are hard for fledglings to swallow, and if they manage, yield little in the way of nutrition. While climate provides one theory for the decline in sandeel availability, another is overfishing by humans. This may well have presented a problem for Sea Eagles in the 19th century, when cod and then herring were target species for commercial fisheries. While these might not necessarily be target species for Sea Eagles, it is probably true that stocks of many fish in inshore waters at that time were seriously depleted.

But even if Sea Eagles lost one or two important fish species, they could well have benefitted from increases in others, whether fish or seabirds. It should be borne in mind that some of these changes in seabird numbers might have come about too late to benefit Sea Eagle populations already in terminal decline. Gulls and Fulmars are species which thrived on discards from commercial fisheries. Eider, too, have increased around Britain and within a century or so had spread from one or two restricted localities to sites along the entire coastline of the north and west. Their increase is one that the Sea Eagle could and would have exploited, to replace any lost prey species.

Competition from other species has sometimes been postulated as one reason for the decline in Britain of the Sea Eagle. Although smaller, the Golden Eagle is more aggressive and there are recorded cases of its supplanting the White-tailed Sea Eagle from its nest sites. Willgohs documented three such instances in Norway, with one of the displaced pairs moved to alternative nests nearby. It is not common for the two species to have much contact, however, one being coastal and the other a bird of inland hills and mountains. Although we now find old Erne nests occupied by Golden Eagles, there is ample evidence to indicate that this is a comparatively recent occurrence, after the Sea Eagle had already long vanished. Some Erne nests have never proved attractive to Golden Eagles and remain vacant to this day. Orkney and Shetland are two such prime locations.

Other than humans, of course, the White-tailed Sea Eagle had no serious predators. Recently, however, it has been shown how the Fulmar can bring about the death of predatory birds, such as Peregrine and, as we saw in Chapter 1, a Sea Eagle released in Fair Isle, by spitting its messy stomach oil at them. This congeals on the predator's plumage and can prove fatal. It may be tempting to postulate that the rapid increase in Fulmars in Britain contributed to the decline in Sea Eagles but this only began when the Sea Eagle was already nearly extinct in Britain; prior to 1878 they bred only on St. Kilda and the subsequent rapid expansion of their breeding range took place after 1900. What's more, Sea Eagles regularly catch and kill adult Fulmars, presumably in flight or as they strive to take off. The Fulmar is attacked from behind, leaving it no chance to retaliate, and any threat comes mostly from young chicks spitting at a predator approaching the nest ledge. Problems from Fulmars might prove to be local and comparatively rare.

There is little evidence, therefore, to suggest that climate, food supply or competition from other species had any significant effects on the decline and subsequent demise of the Sea Eagle in Britain. We are left only with human persecution and it was the contention in my 1983 book that humans alone can indeed be held responsible. Their impact may range from unintentional disturbance at breeding sites during critical phases in the breeding cycle, such as incubation, to the concerted and wilful destruction of eyries, eggs, young and full-grown birds. The case presented may be a circumstantial one but it is, I believe, nonetheless convincing.

We have seen how the species had endured persecution from man for centuries. But only in the 19th century did this reach significant proportions in north and west Britain. The following account will refer only to the Scottish Highlands but it is relevant also to Ireland.

The Highlands were sparsely populated for centuries and it was only during the 18th century that the human population began rapidly to increase. Traditionally, the local economy had been based on beef cattle, with the small native sheep kept mainly for their wool and milk. The stock were herded constantly during the winter months. About 1762 the first Border sheep farmers came north with their hardier breeds of hill sheep, soon to spread across the whole country with the promise of improvement and prosperity. One old man in Caithness was prompted to comment how 'everything was all right in the world except for the Erne and Bonaparte'! (Dr Ian Pennie, pers. comm.). By 1810 huge tracts of grazings were being let out in Caithness, Sutherland and Argyll. In two Sutherland parishes alone, the numbers of sheep increased from 7,840 in 1790 to 21,000 in 1808; by 1820 there were 130,700 sheep in the whole county. The new husbandry was not adopted in the West Highlands until about the 1820s, and not in the Outer Hebrides until the 1840s.

On these sheep farms the local Highland people became redundant and were 'encouraged' (often forcibly) to settle on the coast where they could gain employment gathering seaweed

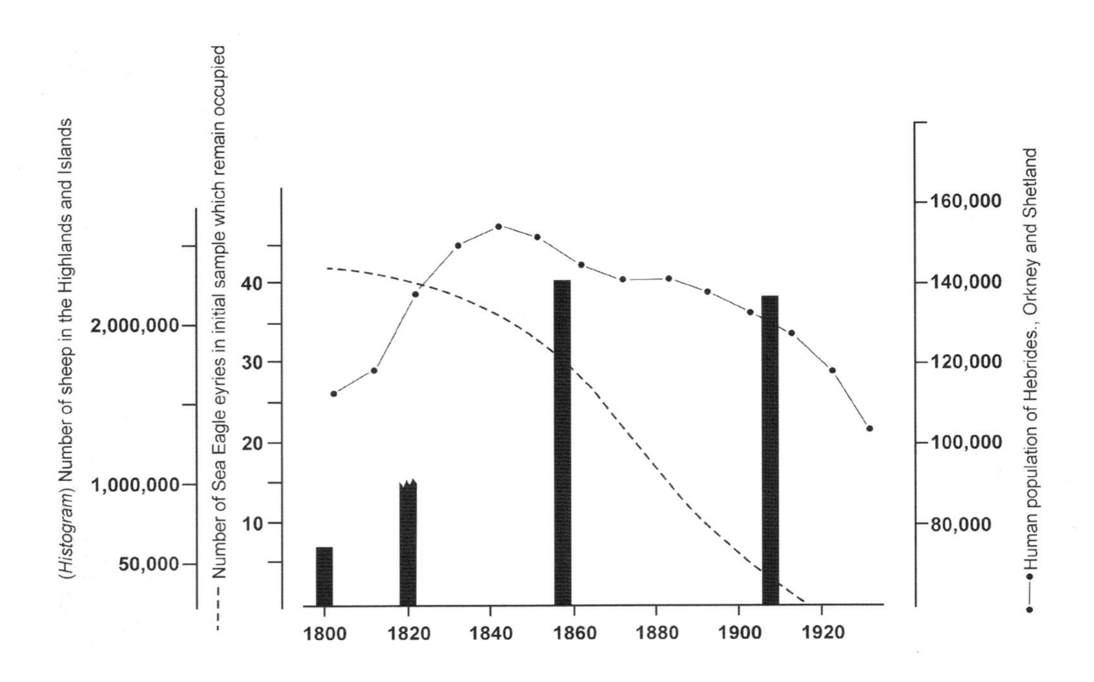

Human population and sheep numbers in the Highlands relative to the decline of the Sea Eagle (from Love, 1983)

or fishing. Their number continued to expand and reached a peak by 1841; in Orkney and Shetland this peak was not reached for two more decades. Thus the increased human population on the coast of the Highlands of Scotland, incidentally in Ireland too, put extra pressure on the White-tailed Sea Eagle, itself of coastal distribution.

By the middle of the 19th century, the notorious 'Highland Clearances' were causing a significant decline in human population, as people were being evicted and transported to the New World, but this allowed no respite for the Sea Eagle. Sheep farming was becoming much less profitable and overstocking had brought about a deterioration in grazing quality. Perhaps seeking a scapegoat for failing husbandry, shepherds harassed predators like the Sea Eagle with renewed enthusiasm and equipped with improved firearms.

Hitherto the flintlock had been efficient only at short range and misfires were frequent. Eagles had to be attracted within range by setting out baits: William Macgillivray described stone huts constructed for the purpose. One old shepherd was said to have shot five eagles in this way in one winter, while another killed three in a single morning. It was, however, a long and tedious business, often for little or no reward.

The instantly-igniting percussion cap was introduced early in the 19th century, improving accuracy and facilitating the shooting of birds in flight. The Sea Eagle possesses an unfortunate habit of circling above its nest, screaming, when someone approaches, thus presenting an easy target. A widowed bird would often quickly find a new mate – and a fresh potential victim. At Fitful Head in Shetland, for instance, one of a pair was shot most winters. The other never failed to find a mate until, ultimately, both eagles were destroyed and the eyrie was finally deserted. As we have said, the motive for these particular killings must remain a mystery, for the locals admitted that this pair had never taken lambs.

I have spent many a long hour scrutinising every old account of Sea Eagles that came to hand. Most make depressing reading, it's true, but one can often glean much about the fortunes, behaviour, breeding and feeding habits of the species. Other, particularly fascinating, products of such research, however, are the occasional insights one gains into the men who perpetrated such acts of destruction. It was a delight to first encounter John Colquhoun of Luss, for instance, along with a local forester called Peter Robertson from Blackmount, whom he often mentions. They both stand out for their depth of experience and their powers of observation with, in Colquhoun's case, an eloquence in putting pen to paper. His *The Moor and the Loch* (1888) is a classic.

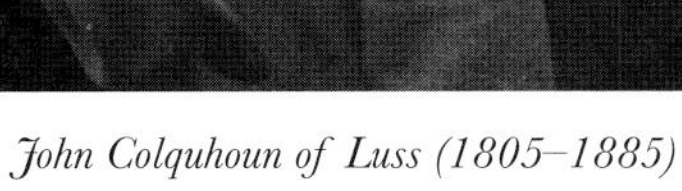

John Colquhoun of Luss (1805–1885)

John Alexander Harvie-Brown (1844–1916)

Harvie-Brown is another such observer who developed an encyclopaedic approach to the wildlife of the Highlands. With a selection of co-authors, his regional *Vertebrate Faunas* are an invaluable resource up to the early years of the 20th century. Harvie-Brown maintained a vast network of correspondents and acquaintances whilst pioneering a questionnaire approach to his fact-gathering. One person he, and others, often quoted is a very old shepherd from Skye called Duncan Macdonald.

Macdonald was in his 92nd year when he was interviewed by the Rev. H. A. Macpherson, Harvie-Brown's co-author for *A Vertebrate Fauna of the North-west Highlands and Skye*. 'The old chap related with gusto his experience of Sea Eagles.' He had been employed for a time as a shepherd in the Isle of Rum, which Macpherson thought was in 1824 or 1825. I think this date is too early, for it was not until 1826 that 300 of Rum's inhabitants were cleared off to Canada, the remaining 100 or so following two years later, by which time the island had become a sheep farm, running 8,000 blackfaces with too few shepherds! Confirming my later date, Macpherson went on to say how it had been 'the year in which my grandfather purchased Eigg'. In fact, we know that old Macpherson bought Eigg in 1827. Only one family was allowed to remain in Rum so the proprietor, Maclean of Coll (or rather his factor), then had to import shepherds for all his sheep. Some came from the mainland, probably the Borders, while others came from Bracadale in Skye. I am convinced that Duncan Macdonald must have been one of the latter.

Duncan Macdonald remained as shepherd in Rum for seven years, during which time he developed an additional interest, indeed an obsession: he is said to have 'killed sundry Sea Eagles'. One year he shot a female at her eyrie. The male then found another mate before he too suffered the same fate. His widow then found a replacement before she too was shot. Not surprisingly, the surviving male discreetly abandoned the site. On another occasion Macdonald killed both adult birds and their three nestlings, a unique bag of five in one day and the only instance of triplets that Macdonald ever encountered.

As we have already mentioned, Macdonald continued his exploits when he returned to his native Skye and, from his little bothy above Glenbrittle, with his double-barrelled muzzle-loader and BB shot, he could kill eagles 'at all times, although spring was best'. Once he shot three Sea Eagles in a single morning, on another occasion a Sea Eagle and seven Ravens all at the same time. Within only a year of his return he had accounted for about 25 eagles in total, after which he continued the business in 'a desultory manner', but, interestingly, he only ever shot one Golden Eagle. At the time Macleod (of Dunvegan?) was offering five shillings for the feet of every dead eagle. But by the time Macdonald came to hand in his 25th Sea Eagle he was rewarded with a pound and a croft in Glendale, and Macleod promptly put a stop to his expensive bounty system. Macdonald's sons John and Kenneth, together with a local keeper named Nicholson, killed many more eagles up to 1876. One account put it at upwards of sixty. It is not surprising, then, that by the time Kenneth moved to Applecross in about 1867, Sea Eagles had already become 'relatively scarce in Skye'.

Immature Sea Eagle feeding on carrion, Rum

By far the simplest, least labour intensive and most effective means of destruction was to employ poison. Harvie-Brown and Macpherson (1904) mentioned how 40 eagles might gather at a carcass; had this contained strychnine or arsenic, the consequences would have been devastating. Understandably we come across relatively few references in the literature to anyone admitting the use of this diabolical mode of slaughter. The eagles which came to nest on Loch Ba in Rannoch, for instance, were poisoned in at least two successive years. In Ireland, Ussher and Warren (1900) astutely recognised how it had been the use of poisons which had brought about the extinction of the Sea Eagle there.

W. H. Hudson (1906) weighed up both sides of the argument and observed how the eagle:

> . . . is regarded by the shepherd as the worst enemy to the flock. But the shepherd has his revenge, for the Erne is a great lover of carrion, and may be easily poisoned.

So much so that John Wolley (1902) was moved to write:

> . . . it is therefore a melancholy reflection that [the White-tailed Sea Eagle] can scarcely exist much longer.

Harvie-Brown and Buckley (1892) went on to plead:

> . . . it would be almost endless and not of much account to enumerate the occurrences of White-tailed Eagles recorded all over the districts of our area as shot, trapped or poisoned in past years. What seems infinitely more to the purpose is to 'put up a little prayer' to the proprietor and shooting tenants of lands formerly and presently occupied by White-tailed Eagles to take active measures for their future protection.

One enlightened tenant on the Park estate in the Outer Hebrides had a provision written into his lease, at his own request, prohibiting eagles being killed without express permission from the Chamberlain of Lewis. Perhaps this was deemed necessary to curb the enthusiasm of the estate employees, one of whom had claimed to have killed 13 White-tailed Sea Eagles in one year. Two local men who contravened this instruction by taking two eaglets from a nest on the estate then had the temerity to offer them back for sale to the proprietor. The outcome of this confrontation, sadly, is not recorded.

In Rum, too, the owner John Bullough apparently afforded a measure of sanctuary to eagles but they were by that time already rare. Harvie-Brown and Macpherson (1904) wrote:

> If they have returned or do return to that island again, the present owner, it is hoped, will be as stern and unflinching a protector of them as his father . . .

One of two Sea Eagle eggs taken from an eyrie in 1907, Kinloch Castle, Isle of Rum

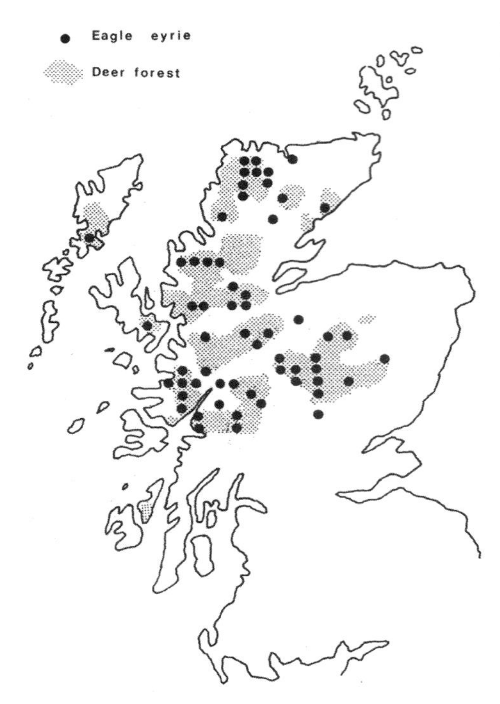

The distribution around 1900 of Golden Eagle eyries in relation to deer forests in the Highlands

His optimism was premature, however, for we know that in 1907, the young heir's gamekeeper took a clutch from an eyrie and shot one of the adults. A pair was shot at the same nest two years later, killings that are all the more regrettable since by then there could not have been more than two or three pairs remaining in the whole of Britain.

Despite his collecting habit, John Alexander Harvie-Brown himself was also a pioneer of conservation. In the 1890s he privately printed a leaflet which he circulated to county councils throughout Scotland, pleading for the better preservation of the Osprey (Love 1982). Sadly, John Colquhoun is often taken to task nowadays for relentlessly harrying Red Kites and for shooting the Ospreys that then nested on the castle ruins of Inchgalbraith, a tiny island in Loch Lomond. It was an action he came to regret wholeheartedly in later life when he often spoke out publicly against the indiscriminate destruction of birds of prey, and campaigned on behalf of Scotland's 'wilderness' areas (John Mitchell, pers. comm.). Clearly these were men ahead of their time.

In general, and as indicated by the map, eagles posed little nuisance on deer forests and were tolerated, while gamekeepers on grouse moors waged an eternal war on all birds and beasts of prey. As early as 1808, the Marquis of Bute insisted that his keepers swore an oath to 'use their best endeavours to destroy all Birds of Prey, etc with their nests, etc... So help me God'. The Duke of Sutherland offered up to ten shillings a piece for eggs and adult eagles, while on a neighbouring Caithness estate the same sum was paid for head or talons of an adult, five shillings for a young bird and two-and-sixpence for each egg. On that estate between 1820 and 1826, a total of 295 eagles (of both species) together with 60 eggs or young were destroyed. In the county of Sutherland from 1831 to 1834 a further 171 eagles and 53 young eagles met a similar fate. These figures include both White-tailed and Golden Eagles. During the years 1837 to 1840, the 'vermin' list of Glengarry estate specified 27 Sea Eagles and 15 Golden Eagles.

Not surprisingly, Sea Eagles were more vulnerable on the coast; on Skye, for instance, practically all of 60 eagles claimed had been Sea Eagles. Of course, one should treat these figures with caution for they are likely to have been inflated, either by the keeper anxious to demonstrate his diligence, by the employer advertising the superiority of his estate management, or possibly even by both. Nonetheless, they do reflect the pressures to which eagles were subjected at that period. Harvie-Brown and Buckley (1892) shrewdly summed it up:

> At one time commoner than the Golden Eagle, the two species have rapidly exchanged places in faunal value. [The Sea Eagle] had not and has not the advantage of the other 'mountain eagle'. It does not receive the fostering care of our deer forests. It is more

> predacious in its daily life. It affects the sea coasts, where it is more open to attacks by sea, and less insured against molestation even by land. It is even more a wanderer; it is more subject to periodic fits of lamb-stealing, and certain old birds, whose 'teeth' get blunted, no doubt do occasionally help themselves to the easiest obtainable supplies. It is even more easily trapped than the Golden Eagle, as being certainly more omnivorous in its diet.

Several years later Harvie-Brown and Macpherson (1904) reassessed the situation:

> I can remember well when the Golden Eagle used to be considered the rarer of our two indigenous Eagles – whether that popular estimate was correct or only an impression. . . The Golden Eagle has taken the place of the Sea Eagle which was formerly the eagle of Skye; though no doubt the two species were often confused with each other.

The writing was already on the wall for the total loss of the Sea Eagle in Britain. John Lamont, gamekeeper on South Uist for 50 years, remembered when Sea Eagles 'nested freely on the island'. By 1919 his successor Donald Guthrie would add:

> I am sorry to say it is now very rare and to all appearance will soon be extinct.

Immature White-tailed Sea Eagles are more prone to wander than the adults. I looked at records in regional ornithologies for Berwickshire, Northumberland, Yorkshire, Wales, Devon and Cornwall. In a sample of 102 leading up to the species' extinction, 80 per cent were immatures. Nearly half of these had been shot. One can presume that most, if not all, were of local British origin and hence added a not insignificant toll to a species already in decline.

Many of these unfortunate wanderers ended up as trophies in glass cases to adorn Victorian drawing rooms, it being increasingly fashionable to display stuffed specimens. Collecting empty eggshells was just as popular. Harvie-Brown was conscious of the 'extravagant prices' offered by collectors. He heard of one notorious egger who, in a period of 25 years up to 1895, had lifted a total of 109 'eagle' eggs; nearly half had originated from only six or seven eyries in Sutherland. We learn how one luckless pair contributed 16 eggs over seven seasons.

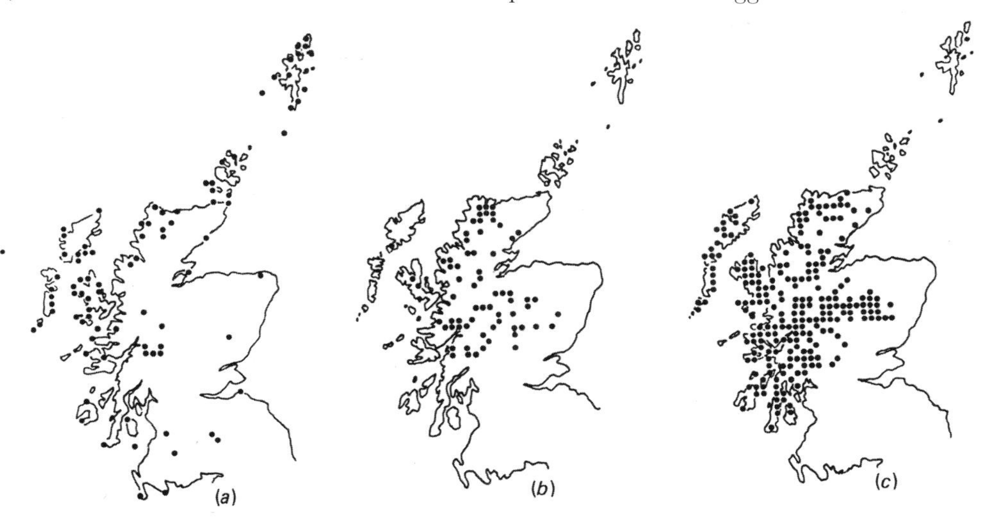

The distribution of Sea Eagles and Golden Eagles in the Highlands during the late 19th century based on various sources (from Love 1983) and compared with the current distribution of Golden Eagles (based on recent Atlas projects)

Only as the White-tailed Sea Eagle became rare did its eggs become especially sought after. The exploits of these obsessive collectors makes villainous reading nowadays. But it should be remembered that they were the naturalists of their time, in days when binoculars were as yet little known and field ornithology was in its infancy. Even the more enlightened Harvie-Brown was an obsessive and, besides eggs and bird skins, he collected semi-precious stones and even bars of toilet soap. Sadly, most of his natural history collection, though fortunately not his library, was lost in a fire at his home in Dunipace, Stirlingshire (Love 1982). His writings, together with those of John Colquhoun, John Wolley and other contemporaries, are some of the only eyewitness accounts we now possess of the White-tailed Sea Eagle living in the wild in Britain.

Because many Sea Eagle sites were conveniently accessible, they were repeatedly plundered. John Wolley (1902) mentioned how one pair nesting on an island in the middle of a loch 'do not always calculate the depth of the water, as there is one place, at least, to which a man can wade.' Other raiders swam, while one inventive Highlander paddled out in two wooden tubs. When Sea Eagles finally ceased to breed at a site in Rannoch, thanks to the regular visit of an egg thief, Wolley was slightly indignant that the thief had been a 'gentleman' – in a boat.

The impressive cliff eyrie at Dunnet Head in Caithness was persistently robbed as well; two eaglets were shot in 1847, Wolley took another two young the following season, and a clutch of eggs in 1849. Not surprisingly, the adults chose in subsequent years to use a less accessible situation elsewhere. Of course, though, these are only the years we know about. Similarly, the eyrie on Whiten Head in Sutherland is recorded to have been robbed in 1849, 1852, 1868 and 1874, and who knows how often in between. The situation of this nest, according to Wolley's informant, was awe-inspiring:

> I cannot take upon me to give a description of the wildness of these rocks, only my hair gets strong when I think of them. After being at this place I always felt some dizziness for two or three days.

A similar situation could be encountered on the towering West Craigs, near the Old Man of Hoy in Orkney, which rose to a perpendicular height of 400 m from the sea below:

> Yet, with the assistance of a short slender rope made of twisted hogs' bristles, did the adventurous climber or rocksman Wooley Thompson traverse the face of this frightful precipice, and for a trifling remuneration brought up the young birds.

A few years ago, Bob McGowan of the National Museum of Scotland sent me a similar account by Simon Ross in the notebooks of the egg collector J. J. Dalgleish. In 1874 the pair raided a nest on the Ardnamurchan Peninsula:

> May 7 . . . now here comes the hard task, the Eagle. When I saw the rock I was rather queer and those that had been with me was taken rather nervous, but I made a start for the nest [on a rope] through rotten rocks and whatnot until I got down about 120 feet and after getting on to a ledge on the rocks I had to turn to the left side about 60 feet . . . and at the end of this ledge was the nest with three eggs in it . . . I took them up and put them in a small basket I had tied on my back. Now here comes the last struggle, a job I'll never forget as long as I will be alive. I got back but I had been cut in more than 20 places with the sharp rocks. I had a new pair of trousers on but by the time I was finish[ed] it was very little I had together of the trousers, and of a pair of boots I had on.

Such exploits were hazardous in the extreme and the courage of these men, if not their motives, of course, has to be admired. As one Shetlander climbed up to a nest he actually caught the unsuspecting female Sea Eagle as she incubated. Despite being 'a very expert and daring fowler' the man was forced to drop the bird on the tricky descent and she was killed. One of John Wolley's climbers took two eggs from a cliff eyrie on Fetlar, but had to drop one while desperately saving his own life. Another Shetlander was less fortunate: when climbing into a nest on Bressay in 1861 'the poor fellow lost his hold, and of course,' – Wolley added somewhat insensitively – 'lost his life.'

In publishing their experiences, Wolley, Colquhoun, Booth and others were very much in the minority. Some might have kept their own personal diaries, a few of which have now come to light. On the whole, though, relatively few egg collectors ever put pen to paper. All that most of them might leave to posterity were faded eggshells gathering dust in drawers and cabinets. Only a proportion of collections survive for reference in museums, and many of these are not even in this country. It is difficult, then, to assess the true extent of egg-collecting in the Highlands or its impact on the Sea Eagle's demise.

I amassed data from as many clutches as I could trace – both Sea Eagles and Golden Eagles – for my 1983 book. The late Desmond Nethersole Thompson – a reformed egger turned renowned and reputable scientist – was especially generous in sharing information with me. All these data have been plotted according to the decade of their collection. There would appear to have been a flurry of activity around the 1850s and 1860s but this could have resulted from the well-documented activities of John Wolley alone. On the other hand, it may be because the Sea Eagle had already gone into decline and its clutches were harder to come by. As the Sea Eagle became scarcer, so the efforts of collectors gained in significance and added momentum to its decline. Shepherds and gamekeepers may have exerted the major impact – certainly so towards the end of the 19th century – and, at the very least, collectors assisted in mopping up some of the last few surviving pairs.

Then, from 1900, a staggering number of Golden Eagle clutches

Sea Eagle egg (left) from Kinloch Castle and Golden Eagle egg taken under licence for analysis, Rum

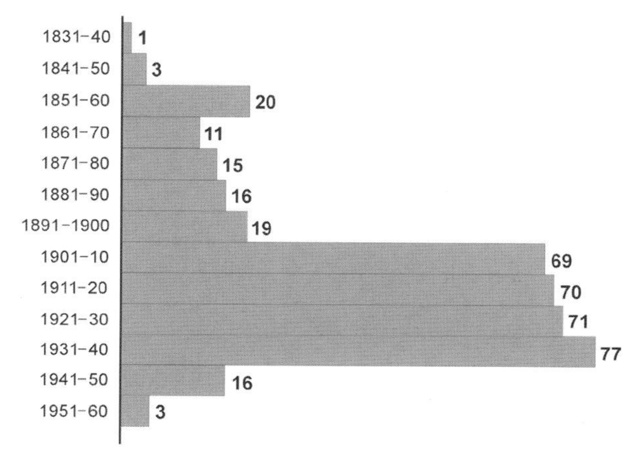

Golden Eagle clutches taken by collectors (from Love 1989)

found their way into collections. These may suggest a sudden surge in the popularity of egg collecting but, perhaps more realistically, reflects both an increased availability of information and an increase in the numbers of Golden Eagles in the Highlands. They contribute a startling 389 clutches altogether, compared with only 51 for the Sea Eagle. Of course one cannot begin to estimate how many clutches have now been lost without trace or which I failed to discover.

Immature Golden Eagle

In this sorry tale of persecution it is apparent that the Golden Eagle must have suffered heavily, if not more heavily than the Sea Eagle, so why should it have survived, still breeding commonly throughout the Highlands? Several factors contribute to the especial vulnerability of the Sea Eagle. The first is that it all too often found itself nesting within easy reach of humans. We have already seen how researchers in Greenland could gain access into 87 per cent of the eyries they visited, and no fewer than 67 per cent were said to be easily accessible. Similar conditions prevail in Iceland and Norway, places where the habitat exhibits many similarities to those found in the Highlands and Islands of Scotland. There are no comparable data published for the Golden Eagle but its eyries, located deep in remote mountainous areas, would seem to be much less accessible.

To a people concentrated on the coast, the nests of a large, noisy and striking bird such as the Sea Eagle would be especially conspicuous and well known. At first, though, Harvie-Brown was more worried about the loss of Golden Eagles, about which less was known, than he was about their larger and more obvious cousin:

> We could say a great deal more about [Golden] eagles in the present area, but refrain. We could mention literally scores of deserted eyries, and a good many of those still occupied, but we cannot see at present that it can answer any useful purpose …

But by 1904 there had been a turnaround of fortunes. Harvie-Brown and Macpherson were registering a recovery of Golden Eagles west of the Skye Cuillin, doubtless because of the persecution pressure to whch Sea Eagles were being subjected. Not only were their eyries familiar and well known, but Sea Eagles were further vulnerable because of their unfortunate habit of flying around overhead, calling loudly, when an intruder is near their nest. This not only advertises their presence, but also makes them particularly exposed to being shot. Golden Eagles prudently disappear from sight, only to return when the danger has passed.

Without doubt, though, the single most important factor contributing to the extermination of the Sea Eagle was its penchant for carrion, leaving it vulnerable to poisoning. This was readily recognised by John Colquhoun (1888):

> A cursory glance will show how much more vulture-shaped both the bill and the body of the Sea Eagle are than those of [the Golden Eagle]. She also partakes of the nature of a vulture in having a less dainty palate than the Golden Eagle and, being not near so quick a game destroyer, is more apt to devour what she does not strike down. Even carrion does not come amiss, especially in winter... she is often fain to content herself with carcasses left upon the inland swamp or cast up by the tide on the shore.

Shepherds and keepers were quick to exploit poison and, understandably, tended to be a bit guarded about this dastardly practice. Harvie-Brown reported a nest in Loch Baa, Rannoch that by 1870 had not been used for several years. One of the adults had been poisoned with strychnine, as later was its replacement and, finally, its mate. Osgood Mackenzie (1928) confessed to a Sea Eagle being found dead in Wester Ross after strychnine eggs had been put out for crows. In the same area, in around 1825, many kites were poisoned by strychnine in a horse carcass.

The protection of lambs was always given as the motive, but in truth every lamb loss was being blamed on Sea Eagles simply because they were seen feeding on the carcasses. No one paused to wonder whether in fact the eagles had actually killed them. There were plenty of dead sheep and lambs lying around. Stocking rates were undoubtedly too high for the quality of the grazings so losses were high, as I was able to show for the Isle of Rum in the 19th century (Love 2001). In the period 1826–1828, 400 people were replaced by no fewer than 8,000 blackface ewes on only 140 sq. km. of some of the wettest and most rugged terrain in the west of Scotland. Even by West Highland standards, it is an exceedingly wet island, with 100 inches or more of rain per annum. Ten per cent or more of adult blackface ewes were being lost each year. Losses among lambs were much higher, the percentage surviving ranging from only 50–70 per cent (67 per cent overall). This could be compared with 63–82 per cent (71 per cent overall) for the Highland area in general. It should be remembered that these latter figures were calculated 50–100 years later than my Rum calculations. By the time Frank Fraser Darling compiled his *West Highland Survey* in 1955, flock sizes throughout the area were probably better adjusted to carrying capacity. Certainly a few decades after the period in which my Rum figure applied, the island was considered overstocked, so blackface numbers had to be reduced from 8,000 to 5,000. Furthermore, in a short-lived experiment with a less hardy breed, nearly a quarter of Cheviot ewes died in their first year, the survivors returning an even lower lambing percentage.

So, without doubt, sheep losses in the Highlands and Islands were very high. Despite, as Colquhoun admitted, the Sea Eagle being 'not near so quick a game destroyer' as the Golden Eagle, they were now the ready scapegoat and an easy target. With so much carrion lying around, Sea Eagles proved all too easy to poison and so shepherds destroyed as many as they could. The Golden Eagle was in all likelihood also targeting poor quality lambs, but the bird was less familiar, less accessible, less ready to come down to carrion, more wary and, indeed, probably rarer than the Sea Eagle.

During the 19th century the Highland (and Irish) people had been cleared from the coast to find employment gathering or burning seaweed (kelp), as fishermen or shepherds. Then, if the inland glens did not prove successful as sheep walks, they became deer forest instead. This provided a good refuge for Golden Eagles since stalkers did not find them a nuisance. The distribution of many Golden Eagle eyries I traced in my researches correlates well to areas of deer forest, while the Sea Eagles were largely coastal.

Only the larger, more mountainous islands of the west, such as Skye, Rum, Mull, Jura and Harris, could have supported Golden Eagles, where they could never have been particularly abundant. Duncan Macdonald, for instance, had shot many eagles in his lifetime on Skye but only one had been a Golden Eagle. Of another 65 known to have been killed in Skye, only three had been Golden Eagles. Again, we find Colquhoun reaffirming how the Sea Eagle was typically a bird of the coast, while to him and others like Wolley, the Golden Eagle was regarded as 'the mountain eagle'.

Even in inland localities such as Rannoch, the two species did not overlap. They were not necessarily excluding one another aggressively but merely exhibiting their own habitat preference, as the ever-observant Colquhoun (1888) pointed out:

> I have enjoyed the rare luxury of seeing both eyries at the same moment, and both queens in undisturbed possession of their thrones. Seldom any collision took place, each having her favourite hunting ground. There was the mountain for the nobler bird, and the morass for her more vulture-shaped neighbour. They sometimes, however, had a battle in the air; but the looser form, the heavier movement and the less daring spirit of the erne made her no match for the mountaineer, who soon drove her screaming to her island...

Colquhoun's Sea Eagle had built on a pine tree growing out from the shore of Lochan na h-Achlaise, while the Golden Eagle reared her brood halfway up a steep cliff to the east of the loch. Colquhoun added that among the exhibits in the Duke of Sutherland's museum, the Sea Eagle had been taken as representative of 'the savage rock coast' whilst the Golden Eagle was displayed as a bird of 'the mountain and the deer forest'.

Deerstalking soon became a fashionable pursuit in the Highlands. In 1790 there were only nine estates used for the purpose but by 1838 – when Scrope's famous treatise on the subject was published – there were 45. In 1912 the number had risen to 203, totalling some 3,584,916 acres. That deerstalkers are traditionally more tolerant of eagles than gamekeepers on grouse moors was cleverly demonstrated by Pat Sandeman (1957). He found that the proportion of lone Golden Eagles on territory, or of one partner of a pair being immature, was greater on sheep ground and grouse moors than in deer forests. This loss of mates was the direct result of persecution. The number of chicks actually fledging on sheep ground and grouse moors was only 0.3 per nesting attempt, compared with 0.6 in deer forests. We can therefore understand why the localities of our Golden Eagle clutches from the last century should correspond so closely with areas exploited as deer forests. During the worst phases of persecution, Golden Eagles in remote mountain areas probably enjoyed ample sanctuary from which to stage a comeback. They gained further respite when gamekeepers were engaged in military duties abroad during World War I (the Lovat Scouts specifically recruited keepers because of their fieldcraft and marksmanship). By that time, though, the Sea Eagle was extinct.

It is interesting that in Ireland both species of eagle were wiped out. Why should the Golden Eagle have succumbed in Ireland but not in Scotland? The answer is simple: both were wiped out in Ireland at the same time, around 1900, at least a decade before World War I. Irish Golden Eagles never had the chance to recover numbers in the respite from persecution which that war, let alone World War II, would have brought.

The depressive effects which can result from human persecution have been exhibited by other raptor species. Many with hooked beak and talons declined and by the early 20th century, seven species had disappeared from Scotland – the Sea Eagle, the Red Kite, the Osprey, the Goshawk, the Hobby, the Marsh Harrier and the Honey Buzzard. It may be significant that the Red Kite displays a particular fondness for carrion which, as with the Sea Eagle, made it an easy victim of poisoning, a significant factor in its downfall.

Ian Newton (1979) was able to demonstrate how the sparrowhawk recovered its former abundance during World War II, when persecution was reduced to a minimum. Buzzards had also been depleted but subsequently were able to build up numbers only in those areas of Britain where gamekeepers were scarce or least active. Furthermore, during World War II, the Peregrine apparently preyed upon carrier pigeons. The Air Ministry encouraged their persecution, thus effectively eliminating the predator from the southwest of Britain, nearest the Channel flightpath from Europe, and considerably reducing it further north. Only when persecution was relaxed after the war did the Peregrine recover its numbers, within the space of only a decade or so.

In summary, it is probable that the White-tailed Sea Eagle was widespread throughout the British Isles. From Anglo-Saxon times – about 1,000 years ago – its habitat in lowland Britain began to diminish slowly. The species had to endure a measure of persecution by man but this seems to have been effective only where the bird was already scarce or had seen its habitat fragmented. The remote coasts of Ireland and northern Scotland could easily have maintained a viable population of Sea Eagles, but several factors which posed a terminal threat came into play in the 19th century. The human population increased and suddenly became more coastal so that interactions between man and Sea Eagles became critical. The bird came into conflict first with sheep farmers and later with sporting interests.

As shepherds and keepers reduced its numbers, so the Sea Eagle became a popular target for egg- and skin-collectors. Large birds of prey are particularly susceptible to persecution primarily because of their long life, meaning that the killing of adults has a bigger impact on the population. Secondly, they have a slow rate of reproduction, producing, if they are lucky, a couple of fledglings per year. Breeding White-tailed Sea Eagles were particularly vulnerable to the depredations of man due to the accessibility of their nests, the comparative ease with which they could be shot at the nest or else be poisoned at carrion. In contrast, the Golden Eagle gained a measure of immunity thanks to its inland habitat, where the human population was thin on the ground, and because it was tolerated where this was used as deer forest. Unlike the Sea Eagle, it survived in sufficient numbers to gain respite from persecution during the two world wars, from which base it could then mount a recovery. Indeed, it now occupies some former Sea Eagle territories on the coast.

It would appear from countless examples, only some of which are reviewed here, that the extent of human persecution directed towards the White-tailed Sea Eagle, as well as other raptors, was sufficiently intense to result in its extermination in Britain. A similar sequence of events had occurred in Europe where, after World War II, the Sea Eagle was impacted by toxic chemicals, as were raptors such as Peregrines, Sparrowhawks and Golden Eagles in this country. This peaked by the 1970s when conservation efforts intensified. Only the unpolluted waters of Norway held a healthy population of Sea Eagles, in clean habitat that was mirrored along Britain's Atlantic shores. These, though, were empty of Sea Eagles. In what could be seen as part of a wider European effort to conserve the species, the stage was set for their return.

Chapter 7
Reintroduction

I hear the eagle bird
With its great feathers spread,
Pulling the blanket back from the east.
How swiftly he flies,
Bearing the sun to the morning.

Iroquois poem

The extinct Haast's Eagle *Harpagornis moorei* from South Island, New Zealand was first described in 1870 by Dr Julius von Haast, a German geologist who founded the Canterbury Museum. Six years earlier he had also helped establish the Canterbury Acclimatisation Society, an august body of settlers who wished to populate the fields, forests and hills of New Zealand with plants and animals from Europe, to serve as a familiar source of food, as sport or amenity, for sundry other utilitarian purposes or for sheer nostalgia. This was soon to become a fashionable pursuit in other colonies round the world, the native flora and fauna having had to live with – or die from – the consequences ever since. European Starlings and House Sparrows now abound throughout North America, and of course rabbits are a huge pest in Australia (Wilson 2004). Examples are legion.

But nowhere demonstrates the disastrous impacts of foreign or 'alien' species introductions more vividly than New Zealand itself, and nowhere there does it better perhaps than a tiny rock in Cook Strait called Stephens Island, less than 3 sq. km. in extent. As soon as a lighthouse was built there, in 1894, the keepers' cats started bringing in specimens of a curious little flightless wren, new to science and by then found nowhere else in the world. The best known culprit, Tibbles, was not solely to blame – if indeed there ever was a cat there called Tibbles. Within a year the wrens were extinct (Wilson 2004). Another classic case is, of course, the flightless Dodo on Mauritius, hunted by sailors as food; there was a similar bird called the Solitaire that lived on nearby Rodriquez. Countless other examples are known from around the world. Flightless birds on islands are naturally especially vulnerable, but not exclusively so, while the cat is just one of many predators responsible.

It is hard to find an alien or invasive species introduction that has proved beneficial, and this phenomenon remains one of the most significant threats to the planet's biodiversity, alongside habitat destruction and over-hunting by humans. But, before anyone now attempts to condemn efforts on behalf of species such as the Sea Eagle, we immediately need to clarify the distinction between introduction and reintroduction. Conservation bodies define the former as 'the release of a species into an area in which it has not occurred before'. The World Conservation Union (formerly the International Union for the Conservation of Nature or IUCN) defines a reintroduction programme as 'an attempt to establish a species in an area that was once part of its historical range, but from which it has been extirpated or become extinct'. We might also add 'as a consequence of human activities'.

There is a third term – restocking – used when the animal is already present but in numbers so reduced that its future survival might depend on topping up. An example of this was the release during the 1970s of Bald Eagles from Alaska into New York State where the species – America's national bird – was on the verge of extinction (Nye 1982). Similar projects followed elsewhere in the USA, by which time the release of captive-bred Peregrine Falcons into the wild had become commonplace (Cade and Temple 1977).

Bald Eagle taking off, British Columbia

All too often, introductions of animals and plants were, or indeed still are, undertaken with no thought for the consequences. Today, most of these senseless acts would be frowned upon and may even be totally illegal. Conservationists, on the other hand, do not undertake reintroductions lightly. Initial guidelines were drawn up by the IUCN, and later by the Worldwide Fund for Nature. In 1979, the UK government set out their own approved procedures based on these guidelines. All are in general agreement. Now too, under its conservation legislation, the European Union positively encourages the reintroduction of lost species, to restore and enhance disappearing habitats and biodiversity.

Firstly, there should be an intensive review of the species past and present and of its new environment to ensure the reintroduction's success. The reasons for the species' original demise should have been rectified, and legal protection put in force. It is obviously advantageous if it has not been too long since the species was first wiped out. Personally I feel that, with some 1,500 years having passed since lynx and bear died out, one should ponder the wisdom of bringing them back. Is the environment really still suitable? With 500 years since the disappearance of the beaver and 250 years or so since the last wolf was slain, the argument for their reintroduction may be a bit stronger, but even if attitudes might have changed, the pressures of over 55 million people in the UK environment must now impose significant impact. It is less than a century since both the Red Kite and the White-tailed Sea Eagle became extinct in Britain, which must be a perfectly appropriate timescale to justify intervention.

It is of course crucial that the environment should still be capable of supporting a viable population yet, at the same time, that that restored population should not then pose a threat to the environment. Importantly, the proposal should nowadays only be undertaken with the support of and due consultation with the local human population, though it has to be said that the initial extermination of species like the Sea Eagle went ahead without any such public consultation! Nowadays it just would not be legal, notwithstanding the continued clandestine and illicit efforts of a tiny minority. There will always remain a certain level of suspicion about reintroductions within any community, which may impede the project, so the ultimate decision to proceed or not must be reached objectively, sensibly and scientifically. If all these considerations are taken into account there should be no need to invoke any exit strategy, though ideally one should exist. We firmly believe that the White-tailed Sea Eagle reintroduction (and that of the Red Kite) fulfilled all these aspirations.

Since the welfare of the individuals concerned is paramount, the capture, transport and release of the animals concerned should be carried out legally, humanely and sensitively, without any adverse long-term effect on the donor population. Finally, the individuals for release should be as genetically close to the original population as possible.

Our Sea Eagle reintroduction story really began in 1959 (Sandeman 1965) and 1968 (Dennis 1968), before any such guidelines had been mooted. Nor had they been agreed and put in place when the first Sea Eagles were released in Rum. The project team were already mindful of good practice but it was not until my book was published in 1983 that the first full appraisal was published, showing that it had already anticipated the IUCN, WWF and UK government recommendations that were by then being agreed.

But back then, who to ask? The government's Nature Conservancy and the local landowner had sanctioned the release of the three Sea Eagles in Glen Etive in 1959. Nine years later, on Fair Isle, the Conservancy, the RSPB, the owner (the National Trust for Scotland) and, most importantly, the islanders who lived and worked there, all gave permission for four more eagles to be released. By 1975 the Nature Conservancy Council (as Nature Conservancy had by then become), as the government's statutory conservation body, had given careful thought to applying the new proposals on the Isle of Rum, their own National Nature Reserve populated only by their staff.

At that time no precedent had yet been set for a full public consultation. There was no real mechanism in place to canvass it, nor indeed had such an approach become as fashionable as it now seems to be. However, the Rum Reintroduction Project was regularly covered across the media. I spent a considerable amount of time with journalists, radio broadcasters and TV crews, or else writing articles for magazines and journals (e.g. Love 1977, Love 1980). In the winters of those early years I also travelled extensively throughout the Highlands and Islands delivering lectures, and attended conferences both at home and abroad. As a result the project was very much in the public domain.

There was also by then huge nationwide support for wildlife conservation, not to mention international initiatives to promote biodiversity. We were not conscious of dissenting voices or criticism of the reintroduction being raised at the time. Indeed, the public seemed quite gripped by the whole concept of the Sea Eagle restoration, being fully aware of the osprey's return. With the Osprey a regular migrant, more and more were turning up in this country, so it was only a matter of time before they started to breed here, quite naturally, in the 1950s. The Osprey population is now well established with over 200 pairs in Scotland and numbers still rising. It has recently been re-established in England and Wales (Dennis 2008). On the other hand the Sea Eagle is much more sedentary, with only some immature birds tending to wander, so it was unlikely to return under its own steam. Since humans had brought about

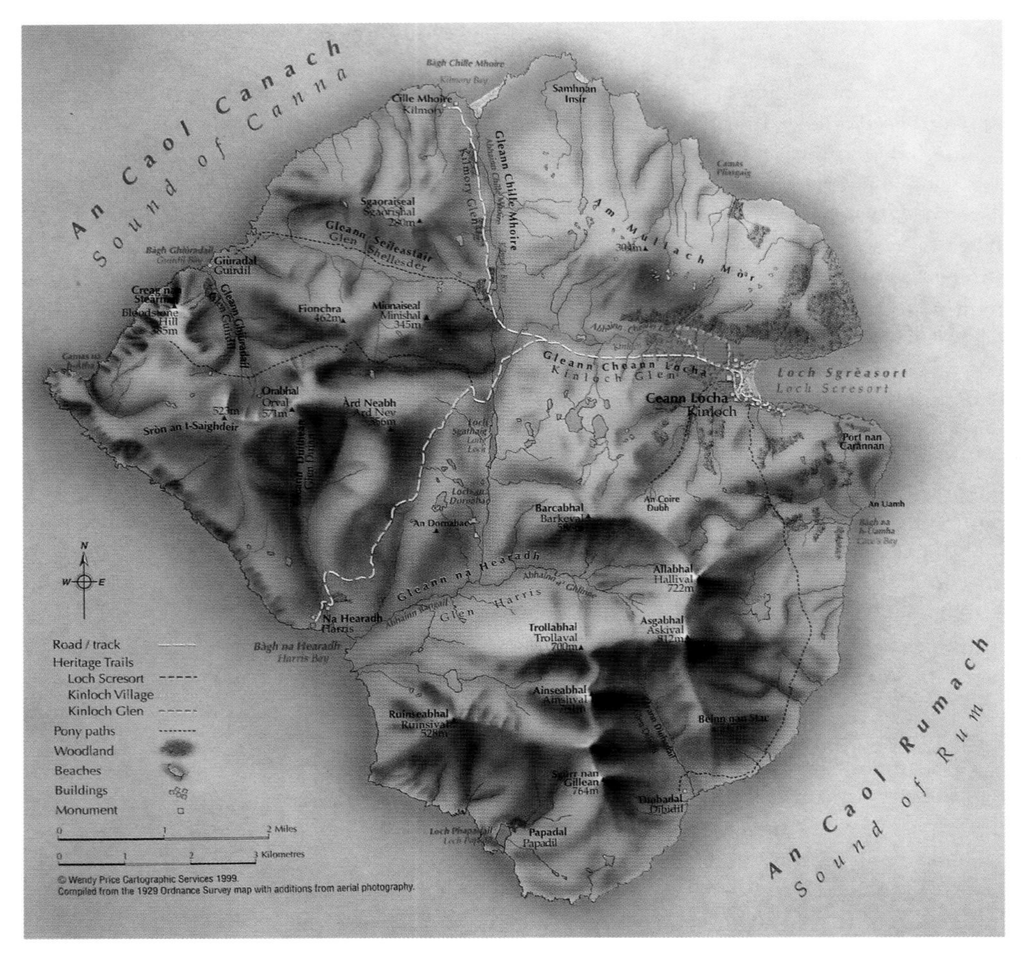

Relief map of the Isle of Rum National Nature Reserve (courtesy of Scottish Natural Heritage)

its extinction here in the first place (without, I repeat, any public consultation), it seemed only fitting that humans should now engineer its return.

In respect of the guidelines that would later be drawn up, I was already pulling together enough documentary evidence to show how widespread and familiar the Sea Eagle had once been across Britain throughout history. Although some habitat loss had occurred, especially in southern Britain, it was clear that the main driver in the extinction of the species in this country had been persecution by humans – trapping, shooting, the destruction and collection of nests, eggs and young, but especially poisoning. In 1954 a comprehensive Protection of Birds Act came into force, and has been updated, improved, revised and renamed many times since. As a consequence, both persecution and egg collecting became much less prevalent though, sadly, have still not been entirely eradicated to this day. We were confident that the habitat was still suitable and that there was still a place for the Sea Eagle.

There was undoubtedly still some suspicion and concern in the public mind, but the sensible assessment was that the killing of lambs by Sea Eagles in the past had been no worse than still went on with Golden Eagles in the Highlands. The Sea Eagle, indeed, is much more of a carrion eater than its cousin, and yet this habit had led its persecutors to conclude, quite wrongly, that it had actually killed all of the lambs, and even the full-grown sheep, upon which it fed. Documentary evidence would indicate that the shooting and poisoning of Sea Eagles to the point of extinction was never justified. It was based purely on emotion, not – as we would demand nowadays – upon scientific evidence. But to strive, even in these more enlightened times, for unanimous backing for the species' return would hinder it ever happening at all. Laying aside the current economic crisis, the proposed reintroduction today in southeast England has been put on hold because of a powerful, vociferous, and – it has to be said – misinformed minority lobby by farmers and estate owners.

It is certainly a long time since Sea Eagles were common in England but up to 1958 some 338 occurrences are documented, mostly in autumn and winter, mostly on the east coast and almost all involving immature birds. Lamentably, many of these ended up being shot. Surprisingly, in the period since 1958, only 24 have been recorded, but by that date Sea Eagles were under considerable pressure abroad from threats such as pollution. Sadly, even in this day and age, the one bird which appeared in Norfolk in 1984 was shot; it had been ringed as a nestling in Schleswig-Holstein only the year before. Another seen in Hampshire during the 2007/8 winter had come from Finland, with yet another Finnish colour-ringed individual sighted in March 2004 at Loch of Strathbeg, a few miles south of Fraserburgh in northeast Scotland. It is interesting to note that a young Sea Eagle colour-ringed in Norway in 2011 was sighted at the other end of the country, in Unst, Shetland, during December 2012). Wandering immatures tend to gravitate back to their natal area as they near maturity. Not only would potential natural colonists have to stay put and survive for at least five years, but they would also have to find a suitable mate, a safe territory and a nest site.

Sightings in south and east England are likely to become more frequent as immature birds wander west, now that populations in the Baltic and Low Countries are increasing. A pair of Sea Eagles arrived to breed in Holland as recently as 2006, and there are now several others nesting. Despite farmers' fears nesting attempts are bound to take place in southern England in the not-too-distant future.

Sub-adult Sea Eagle, Rum

Back in 1975, though, when the Baltic populations were still at a low ebb (Helander 1975), the chances of Sea Eagles re-establishing naturally in this country were considered remote. It seemed then that time was of the essence for Sea Eagles in northern Europe, if we were to establish a viable, reserve population in Britain.

As long ago as 1904 Harvie-Brown and Macpherson had commented:

> Our natural sense of regret at the extermination of the [sea] eagle in Skye is tempered by the consideration that the wide range it enjoys renders it almost impossible that it could ever become a lost species, since it ranges from Greenland to Siberia, and from the fjords of Norway to the forests of Hungary.

While this may yet be true, it does not mean that the species is on a safe footing. Certainly, the White-tailed Sea Eagle does (or at least did) enjoy an extensive Palaearctic distribution from the edge of the polar region south to the deserts of Israel, Egypt and the Middle East, and from Greenland right across Europe and Asia to the Bering Straits (see Appendix II). Within this impressive span it copes with a huge range of climatic conditions, habitats and human interactions. While a great deal is now known of the species in Western Europe, there is still much to be learnt about its status throughout most of its range to the east. We simply do not know, then, how secure are the Sea Eagle's prospects.

At the end of the 19th century in Western Europe, the White-tailed Sea Eagle bred along the entire coast of Norway from the Swedish border in the south to the Finnish, now Russian, border in the north. A noticeable decline then took place, probably due mainly to human activity, predominantly persecution by hunting, trapping, nest robbing and the use of poisoned baits. And such impacts were not felt just in Norway but throughout Western Europe where, indeed, the situation would prove much worse.

One way to stem the decline of endangered wildlife is to give it legal protection and so minimise human persecution. It would not be until 1968 that Norway extended full legal protection to their Sea Eagles, one of the last countries in Western Europe to do so. It was probably influenced in this decision by the interest shown in Scotland in its reintroduction. Italy followed Norway's lead three years later, as did Greece in 1973. One of the first countries to protect the Sea Eagle had been Iceland in 1913, when their population was on the verge of extinction. Sweden and Finland followed suit seven years later, with Denmark following in 1931. The laws in Germany were complex, to say the least. In Prussia, for instance, it was permitted to destroy Sea Eagles and Ospreys but not by using firearms. The White-tailed Sea Eagles were accorded full protection in Pomerania in 1922 and in Mecklenburg four years later, but not until 1935 did these laws apply throughout what became East Germany. In Germany, even today, all birds of prey are still considered game species, although conveniently the 1936 Act omitted to specify an open season when they might legally be shot. Yet hunting laws still override conservation legislation, despite EU Conservation Directives demanding that this confused situation be rectified forthwith to give full protection to all birds of prey in Germany and bring it into line with the rest of Europe (Bijleveld 1974).

It is one matter to pass laws, of course, yet quite another to enforce them. Between 1946 and 1972 a total of 194 Sea Eagles were picked up dead in East Germany and nearly half of them had been shot (Oehme 1977). In countries like Iceland and Greenland, until comparatively recently, eagles continued to die in traps set for Arctic foxes, while poisoning wolves in Bulgaria and Romania, for example, continued to account for many more. Paradoxically, as the wolves disappeared, the eagles began to suffer further through the loss of carrion from wolf kills. An alternative approach to safeguarding rare and declining species was to stem the destruction of their habitat and to set up nature reserves, both of which proved effective up to a point. But by the 1950s and 1960s it was evident that a new threat had appeared against which no amount of legal protection could prove effective – pollution from toxic chemicals (Newton 1979).

First the organochlorines DDT and dieldrin were routinely used as insecticides and sheep dips, while alkyl mercury compounds were widely applied as fungicides. The stability and persistence of these chemicals resulted in their accumulating in increasing dosages up each level of the food chain until predators, such as peregrines, sparrowhawks and eagles at the top, accumulated levels that either impaired their breeding success or else proved lethal. Long after the use of these toxic chemicals was banned, they continued to leach out of the ecosystem into water courses and ultimately into the sea. The Sea Eagle was especially vulnerable but with the passage of time, conservation efforts in Sweden and Finland in particular (Helander 1981, Hario 1981) began to pay dividends. With the provision of clean, uncontaminated food, for instance, Sea Eagle populations have since recovered significantly (see Appendix II).

Back in 1975, though, with this climate of concern throughout Europe, the clean waters along Norway's Atlantic coasts were still offering vital refuge for Sea Eagles. It was soon realised that similar conditions prevailed around the British Isles where the species had long been extinct and where it was deemed that it could still thrive. A reintroduction project in Scotland could play a significant role in the survival of the species in Western Europe. Fortunately, since then – and in the nick of time – conservation efforts have borne fruit, so much so that in 2005 the IUCN Red List status of the Sea Eagle was upgraded from Near Threatened to Least Concern. It is nice to think that the British reintroduction effort may have played its part in that decision.

At this point we should discuss the options that were considered when deciding how the reintroduction should be carried out. The most immediate results might be expected by planting adult birds, or already established pairs, into vacant territories. It is not easy, although not impossible, to capture adults – especially pairs – in the wild. Neither can it be assumed that the pair bond or the urge to breed would remain intact after such trauma, nor that the pair will choose to settle and breed in their new home. Furthermore, to consider instead the release of adults that have been born and raised captive for several years is of questionable merit. There will often be doubt as to their provenance, while imprinting and their unfamiliarity with a wild existence could prove insurmountable.

Two downy Golden Eagle chicks in an eyrie in Benbecula

Two downy Sea Eagle chicks, Nordland, Norway (Harald Misund)

An alternative approach might be to remove eggs or young from captive or wild pairs and to place them with a suitable foster species such as the Golden Eagle, which is widespread in the Highlands, in places even nesting on the very coastal ledges formerly occupied by Sea Eagles The main problem would be imprinting upon the wrong (foster) species. Using twins rather than single chicks might overcome this but, without disruptive supplementary feeding, it is unlikely that golden eagles in the west of Scotland would be able to rear twin young. The precious eggs, and eaglets, would still remain vulnerable to all the vicissitudes of incubation and fledging, so a successful outcome could not be guaranteed.

From egg to fledgling - a rare opportunity during the project in Rum to see an empty egg from Kinloch Castle, a four-week-old chick and one nearly fledged!

The north and west of Rum from the air

A simpler approach could be to install well-grown eaglets into an easily accessible artificial nest and for humans to act as foster parents, much as has been done with peregrines in the US. Appropriate measures could be taken to reduce the risk of imprinting, with the young allowed to fledge normally, yet still provided with food nearby until independent. This is essentially a technique called 'hacking' which falconers have used for centuries to give young hawks some hunting experience in the wild before being recaptured to fly at the fist. It is this very approach, indeed, modified to suit our particular requirements and conditions, which has been used to reintroduce Sea Eagles to Britain, first in Glen Etive, then Fair Isle and Rum, through two further phases to the present day.

We needed to identify a sufficiently rich source of Sea Eagles to allow a release into Britain without risking any prejudice to the donor population. At that time only Norway was suitable, and it offered the added advantage of being the closest population geographically and, therefore, genetically. Most importantly, the habitat in Norway, especially in Nordland in the north of the country, is remarkably similar to that on offer in Scotland. It could confidently be assumed that the birds would adapt to the wild quickly and easily.

We were indeed fortunate that the Norwegian government not only supported the reintroduction project from the outset but were willing to donate young birds (no money ever exchanged hands). They also facilitated all the necessary export/CITES/veterinary licences. Local landowners in Nordland generously allowed some of their eaglets to be collected each year and we could not have worked with a more dedicated contact and field worker in Norway than Harald Misund, a captain in the Norwegian Air Force in Bodø. Some of his expenses each season were met by WWF (now the Worldwide Fund for Nature). Indeed, the Norwegian authorities have continued to co-operate over many years, hosting our own RAF Nimrod aircraft at Bodø and even offering transport from their own Air Force when needed. We had approached the Ministry of Defence and RAF Kinloss in Moray, from where Nimrod reconnaissance aircraft flew regular exercises into Bodø and could offer a flight across the North Sea.

The next step was to identify a venue for release in Scotland. St. Kilda, as initially mooted by Morton Boyd and Pat Sandeman all those years ago, was ruled out as too remote, bringing huge logistical problems. We ended the first chapter with four birds released on Fair Isle dispersing and eventually dying. It was obvious that Fair Isle was small (810 hectares, rising to only 217 metres) and isolated (35 km from either Orkney or Shetland). If the released eaglets wandered, there was a reduced chance of finding their way back or indeed meeting up and pairing once they had matured. It also had a substantial population of Fulmars, whose oil-spitting defence proved one Sea Eagle's undoing.

On the other hand, the Isle of Rum was a National Nature Reserve owned by the Nature Conservancy Council and staffed by about 8–10 families. At the time, public access was limited due to conservation and research priorities, although it had never been prohibited. Climbers, walkers, naturalists and research students all came to enjoy the wilderness that Rum had to offer. Four or five visits were made weekly by the passenger ferryboat from Mallaig, transferring ashore, before the new pier was built, by the small flit boat. This, together with a shortage of overnight accommodation, had always imposed limits so visitor pressure was never high. In recent years however, public access, especially by day trippers, has been actively encouraged.

Rum is extensive: 10,600 hectares rising to several mountain peaks, each about 800 metres in altitude, which present a conspicuous silhouette from all directions. It lies only 24 km from the mainland and lies in the centre not only of the Inner Hebrides but of the Sea Eagle's former range in Scotland. I was amazed how similar Rum proved to be to the island of Landegode, one of the sites just off Bodø where we were to collect Sea Eagle chicks for Scotland. The habitat we had to offer was amazingly reminiscent of their homeland.

The last known breeding pair in Scotland had been in 1916 on Skye, only 13 km away, with the last pair on Rum itself less than a decade prior to that. If Rum did not detain the released birds, the island was well placed to act as a springboard for their colonising the west Highlands and Islands.

The initial scheme to reintroduce Sea Eagles to Britain was devised by Dr Ian Newton (of NCC, and later the Institute of Terrestrial Ecology) at the request of a senior colleague, Dr Derek Ratcliffe, Chief Scientist of NCC. Ian had long nurtured the dream of seeing Sea Eagles back in Scotland and remembers:

> I was greatly relieved when, after hearing arguments for and against from the various stakeholder bodies, the relevant committee approved my proposal for a fresh reintroduction attempt involving a more sustained programme with a larger number of birds. However, as a result of mainly agricultural concerns, the reintroduction was constrained to take place on Rum, an island reserve entirely owned and managed by the Nature Conservancy Council in Scotland (now Scottish Natural Heritage). Those of us closely involved at this time appreciated that Rum was not the best place for a reintroduction; the climate was exceptionally wet, and the food supply relatively poor for most of the year. But we felt that, once the imported young birds had achieved independence, they would disperse off the island in search of better habitat elsewhere, which indeed is what they did.

I am aware too that Civil Service mandarins were questioning the viability of Rum as a National Nature Reserve, being unique but expensive in requiring the maintenance of a permanent community to service it. It had already delivered a pioneering study into red deer behaviour and management and was active in an innovative forestry scheme to recreate

Dr Ian Newton and the late Dr Derek Ratcliffe

Roy Dennis is also involved in the Red Kite Reintroduction, as well as that of his beloved ospreys and many other innovative projects

A Golden Eagle eyrie in the mountains of Rum with the eaglet surrounded by Manx Shearwater prey

the native woodland cover, both of which continue to this day. NCC in Scotland felt that the Sea Eagle Reintroduction would add another high-profile feather to its cap. So Martin Ball, Deputy Regional Officer based in Inverness, together with Dr John Morton Boyd, Scottish Director of NCC, pulled together the logistics, with constant specialist advice from Roy Dennis.

Although Roy had now moved off Fair Isle – where George Waterston had staged his pioneering effort – and become Highland Officer for the RSPB, he did experience difficulty in enthusing his new employers about the Rum reintroduction. Changed days! I had just finished postgraduate studies at Aberdeen University Zoology Department's Culterty Field Station on the Ythan Estuary, and returned home to Inverness. Through my friendship with Roy, Martin engaged me to manage the project on a six-week contract, based in Rum, where I liaised with the then Chief Warden Peter Corkhill and his staff.

Rum's extensive rocky coastline supports colonies of eider, shag, auks and gulls, all potential Sea Eagle prey. Fulmars, on the other hand, are scarce, with 400 or so pairs having a localised distribution, mainly on the southeast cliffs. Rum is also unique in holding a mountain-top colony of Manx Shearwaters, which in 1975 was reckoned to total in excess of 100,000 pairs (they have since declined significantly). Despite visiting their nest burrows only at night, some shearwaters fall prey to Golden Eagles so would be equally available to Sea Eagles. As well as gulls, otters are plentiful around the island's shores, and from them the Sea Eagles could pirate fish. There are no rabbits, hares or sheep, but carrion abounds from some 1,500 red deer and over 200 feral goats (Love 1980a, Love 2001). And, of course, the waters around the island are as productive as anywhere in the Hebrides: the local fishing boats and the Mallaig fish merchant Andy Race were to prove generous in delivering waste fish to the island for me, and later in throwing scraps overboard for the released birds.

I identified suitable locations in Rum for the cages, well away from human contact but still accessible by Land Rover via what passed on the island for roads. I never reached fourth gear and the time taken for me to cross the island eventually doubled as the track deteriorated further. The ultimate location had also to respect Government regulations, which demanded that the eagles be quarantined for six weeks at least 8 km from domestic poultry. I selected two sites near the end of the rough track to Harris, a remote glen on the southwest coast of Rum. Overlooking the sea, the situation offered ample opportunities for the released eagles to scavenge and to tap into the seabird colonies nearby.

The cages were constructed by NCC estate staff on the island and were based on the design used by Roy Dennis on Fair Isle. They were of chicken wire or chain-link mesh stretched across a timber frame and contained two compartments, each 4 metres square and 2 metres high, large enough to permit the occupants to exercise freely but sufficiently restricted to prevent the eagles from damaging themselves against the mesh. Each was provided with a log perch and an open-fronted wooden shelter a metre square. The birds were to be kept as pairs in adjacent pens while my visits to feed them were to be brief and infrequent to avoid them imprinting on anything other than each other. Two cages were constructed in 1975 but when we realised that the project would continue and that we might get more birds each year, two more were built, with an additonal tethering site nearby.

By June 1975, amazingly, everything was in place for the arrival of the first Norwegian eaglets, a delivery offered mostly free of charge or with costs being borne in-house, with the only major (although not large!) expense being my salary for supervising the project in Rum. Initially I was employed for six weeks, with contracts being renewed regularly, although often at

Estate staff constructing cages, Rum

The late Dr Johan Fr. Willgohs aboard Larus *in 1987*

– or even after – the last minute. I ended up living in Rum for nearly ten years, longer than we had ever anticipated the project would last when we first started.

Due to his expertise and his help in the past on Fair Isle, Johan Willgohs was again approached to be part of this project. Born in 1915, Johan had been active in the Norwegian resistance as a youth during the war. He was eventually arrested by the Germans for helping run an illegal newspaper. I was told that he gave himself up so that his elderly father could escape internment. While incarcerated, Johan was given the task of cleaning his captors' offices. When they were empty he would stand guard outside so that his fellow prisoners could creep in and listen to radio broadcasts from London. After the war he resumed his studies and became interested in seabirds, Eagle Owls and eventually Sea Eagles.

For his PhD thesis, Johan observed Sea Eagle behaviour from hides carefully constructed overlooking eyries. He worked out that the eagles could count to three and he always tried to be seen into his hide by three other people so that when they left, the eagles were content that the hide was unoccupied. Once he had camouflaged the hide so well that he only found it when he fell through the roof. Carefully documenting all he could discover about Sea Eagle behaviour and habits, learning both from his own extensive field work and from the efforts of others, he published his PhD thesis in English in 1963. He was able to confirm that Norwegian Sea Eagles ate mostly fish and birds and were of little danger to livestock.

Between 1956 and 1960, while still based at Bergen University, Johan had organised a census of breeding Sea Eagles in Norway. He estimated a population of some 350 pairs. Soon afterwards he had a visit from George and Irene Waterston and offered to collect four eaglets, to be released on Fair Isle in 1968. That year full legal protection appeared to have checked the decline in Norway, and would eventually even permit local increases. His passion for Sea Eagles continued all his life and, in 1991, just a few years before his death from cancer, he visited me in Rum – but not before I was able to visit him at his island cabin near Bergen.

Johan's population estimate is now acknowledged to be a serious underestimate. Projekt Havørn, the Norwegian Sea Eagle Project, began in 1974 to encourage co-operation between amateur ornithologists, nature conservation organisations, bird societies and scientific institutions, aiming to protect and survey the species throughout Norway (Folkestad 2003). The Worldwide Fund for Nature (WWF) provided funding, while the Norwegian Ornithological Society organised fieldwork. In 1975 the project was incorporated into WWF's international Sea Eagle Project No. 972. Since 1986 the bulk of the funding has been assumed by Norway's Directorate for Nature Conservation.

The project's first nationwide census was undertaken between 1974 and 1976. It revealed about 450 pairs, an increase due, no doubt, to better coverage but also, it is thought, reflecting a genuine recovery in numbers. The true total is still thought to have been underestimated, however, and a reappraisal of the data by Alv Ottar Folkestad and his colleagues indicates that there may have been 700–800 pairs at that time. Using these two surveys as a baseline, monitoring continued. By the millennium, breeding was confirmed at 1,283 different nesting sites and was strongly suspected at a further 381 sites. Furthermore, the census also logged some 280–375 sites where breeding might possibly be occurring. This situation continues to improve.

The bulk of the Sea Eagle population was to be found in northern Norway, with nearly 60 per cent in the three northernmost counties of Nordland, Troms and Finnmark, more than a third of these in Nordland alone. Here the coastal environment remains unpolluted and the eagles' breeding success is high. Another 30 per cent of the population occurs in the central Norwegian counties of Nord-Trøndelag, Sør-Trøndelag and Møre og Romsdal with the remainder lying south of 62° N where the main expansion is now taking place. In the south a few pairs may venture further inland, with a few 10 km or more from the coast (Folkestad 2003).

The Smøla archipelago (277 sq. km) in Møre og Romsdal has a population of some 65–70 pairs of Sea Eagles and although it is tricky to work out the concentration of Sea Eagles on such a convoluted coastline, peppered with countless islands, this is the densest population of the species documented anywhere in Norway or, indeed, anywhere in the world. This area does have its dangers, however, for, since the first of 68 wind turbines began to be constructed on the island of Smøla in 2005, no fewer than 38 Sea Eagles (up to June 2010) have been picked up dead. Further research to reduce this horrific statistic is now urgently underway (Folkestad 2003a).

Initially our priority was to get the project up and running and to see how things developed from there. Johan Willgohs would undertake a Sea Eagle survey along the Norwegian coast on which Martin Ball would accompany him for a couple of weeks. On 7 May, Johan's motor launch *Larus* was stowed aboard a coastal cargo vessel from Bergen to the Isle of Vikna, thence to survey northwards as far as Bronnøysund. From there, on 24 May, Martin had to fly home, leaving Johan and his companion to continue northwards, eventually to meet me in Bodø where I would then join him for two weeks in June 1975.

Smøla with its 68 wind turbines in Møre og Romsdal, Norway

Martin Ball of NCC handled all the organisational tasks from his office in Inverness, but often visited both Rum and Norway

Bodø harbour in June 2010

Bodø, lying just north of the Arctic Circle at 67° 16' N, is the capital of Nordland with a population of some 36,000. It arose as a major fishing port and soon became an important stop for the Norwegian coastal ferry ships, collectively known as the *Hurtigruten*, both northbound and southbound, as well as a terminal for many other smaller ferries to the offshore islands. It is also where the railway from southern Norway finally terminates. During World War II, after Germany had invaded the south of Norway in 1940, Allied troops – Norwegians, Poles, French and British – tried to hold Narvik, the railhead for iron ore westwards out of Sweden. The British built a makeshift airport at Bodø to provide air support for the Royal Navy but soon afterwards abandoned northern Norway to German occupation. The town was bombed flat by the Germans, leaving half the population (then over 6,000 people) homeless. When it was rebuilt after the war, Bodø became one of the country's best-planned towns. The new civil airport, opened in 1952, also served as Norway's largest military air base. During the Cold War, its NATO duties involved the policing of the northern frontier with Russia, and it served as the operations base for the U-2 spyplanes. One of them, famously piloted by Gary Powers, was shot down by the Russians on its return flight from Pakistan to Bodø. A U-2 hangs in the town's extensive Aviation Museum, opened in 1994.

When I first visited Bodø, a squadron of ageing Starfighters took off each morning with a deafening roar, passing right over the municipal campsite. They were not known for their airworthiness so, none too soon, were replaced by F16s. Norwegian Air Force search and rescue helicopters were also based here. RAF Kinloss has now closed and, sadly, Bodø Air Base is now threatened. The reintroduction of Sea Eagles to Scotland owes much, of course, to the Nimrods of 120 Squadron, RAF Kinloss, but also to the Norwegian Air Force at Bodø. Their part in the project was greatly facilitated by Captain Harald Misund. Stationed at Bodø since 1960, Harald had become an expert in the Sea Eagles breeding locally, and from 1976 would take over collection duties for us, firstly for Rum (Phase I) and then during Phase II in Wester Ross.

On Saturday, 1 June 1975 I landed at Bodø airport and took a taxi to the harbour, where alongside commercial vessels lay hundreds of yachts and other small launches. I had a photograph that Martin had given me so spent an hour or two trying to match it to a single motor launch called *Larus*, only to confirm that it was not there but was steaming quietly in from the open sea. Next morning we cruised back out again into the myriad of small islands and skerries that are home to Nordland's Sea Eagles. Several of the main island groups supported small communities of fishermen and farmers, a few of whom supplemented their incomes by collecting eider down and cloudberries, even hunting otters for skins. They all shared a lively interest in wildlife and several were valuable contacts for Johan in his quest for eagles.

Johan Willgohs with his motor launch Larus *in 1987*

Our first port of call was the inhabited archipelago of Helligvaer, with huge wooden racks of drying cod (stokfisk), where the local councillor came out with us for the afternoon to point out several Sea Eagle eyries. Next day we explored the beautiful islands of Karlsøyvaer, a nature reserve lying just offshore from the historic trading post of Kjerringøy. It was here that the Nobel prize-winning novelist Knut Hamsun based some of his books. Karlsøyvaer itself had been recently abandoned by its dwindling human population, although some still used their old homes as summer bases while fishing. Here we found a few occupied eyries as well as a roost of nearly twenty non-breeding Sea Eagles, immatures and sub-adults. Below the guano-smeared birch trees grazed sheep, quite unconcerned.

Colin Crooke overlooking Karlsøyvaer Nature Reserve from a Sea Eagle communal roost

White-tailed Sea Eagle twins in a large eyrie in a low tree in Karlsøyvaer Nature Reserve, a few hundred metres from a Sea Eagle communal roost

Sjark boat moored off Maløy pier, Karlsøyvaer, 1975

Further north, we inspected some other likely islands and at Maløy the villagers directed us to some occupied eagle sites. Our main contact was not at home, though, but out hunting otters which, at that time, still fetched 50–100 kroner; about twenty pelts were hanging out to dry by his barn. Johan explained how otters are important to hungry Sea Eagles, who sit and wait for one to come ashore with a fish. Hardly has the beast had a chance to tuck in to its meal than the eagle swoops down to steal it, lumpsuckers being a particular favourite. Our contact's boat was lying at its mooring, a small, single-man vessel called a sjark boat, with varnished wooden hull and white stripe along the gunwhale.

We met another of Johan's contacts back in the thriving fishing community on Helligvaer and he took us to some remarkably accessible eyries on the flat islands close to the main village. *Larus* continued north to 68° where we looked at a Sea Eagle eyrie on a sheer cliff face before calling at a little village called Forsan to refuel and stock up on bread and supplies. The weather turning wet, we turned south again, closer to Bodø, to ensure we could collect our four eaglets and make our rendezvous with the RAF Nimrod at the airbase. On our way we sheltered off the north end of the island called Landegode and then again at another group of some 60 islands called Bliksvaer. This had been declared a nature reserve in 1969, with the consent of the 20 or so fishing families who lived there at that time. Since then the community has dwindled in size and the very last two to quit – some two decades later – would be the warden and fisherman Hallstein Christiansen and his elderly mother.

They were extremely hospitable with an amazing story to tell. As the Allies prepared to launch an assault on Narvik in an attempt to stem the German supply of iron ore from Sweden, and to protect their own Arctic supply run to and from Russia, it was here at Bliksvaer on 18 May 1940 that the British cruiser HMS *Effingham* ran aground. She was less than an hour away from Bodø. All 1,000 Allied troops (South Wales Borderers) and 700 crew aboard were rescued. The British then scuttled the ship rather than let it fall into enemy hands. With the Allies under pressure at Dunkirk it was decided to abandon northern Norway once and for all, leaving the German bombers to blitz the towns of Bodø and then Narvik to crush any further resistance. The Bliksvaer islanders salvaged as much as they could from the wreck of HMS *Effingham*, only to be raided by the Germans who confiscated everything. Or so they thought, Hallstein's father hid a hand gun from the wreck and when Hallstein had me to lunch one day in July1985, his mother laid it out on a neatly starched Royal Navy tablecloth. I also have in my house an umbrella stand which is a polished brass six-inch shell case from the ship, a gift from Harald. A few years ago I was amazed to come across one of HMS *Effingham's* torpedoes in a museum in the Maldives, the ship having served in the Indian Ocean between the wars.

Hallstein Christiansen, warden at Bliksvaer (Harald Misund)

The next day, in his sjark boat, Hallstein took Johan and me to a couple of eyries on Bliksvaer where we ringed the eaglets and viewed the Cormorant colony, one of the few in northern Norway. On 22 June we finally made a run out of our idyllic little anchorage in Bliksvaer and braved the open sea to head south as far as Fleinvaer, but continuing stormy weather next morning forced us to retreat north again to the shelter of Bodø. With both time and weather closing in, we were becoming anxious so secured permissions to take two well-grown eaglets from a nest in Karlsøyvaer to

Sea Eagle chick at Bliksvaer

With two chicks hatching a few days apart, the first chick, especially if it is a female, is bigger than the second, Nordland

Air France Concorde with Nimrod at Bodø in June 1975

Johan at Sea Eagle nest, Bliksvaer

the north, and the other two from Steinsvaer just out of Bodø. In future years Harald would always take only a single chick from broods of two, leaving the parents with the other to rear as normal.

Pulling alongside in Bodø by 11 o'clock that night, Johan went off to link with the NAF and returned with two airmen who were to take me back to the airbase with the eagles. I said my sad farewells to *Larus* before being whisked off for VIP treatment at the Air Base. The Mess was opened to get me a Coke and a snack, and I was given a room in the Visitors' quarters, with the unfortunate name of 'Cell' Block H. The eagles were accommodated in the shower next door, once I'd been assured that they would have exclusive use. I spread newspaper liberally across the floor and, having fed all the birds back at the boat with Johan, released the eaglets from their boxes and gratefully hit the sack myself.

Next door I had heard the eagles' talons clanking around the marble floor all night, the birds beating their wings constantly to exercise and to rid themselves of itchy down. Unfortunately, it was only now that I realised that the shower was also, after all, for an airman in the adjacent room. When I apologised over breakfast he merely shrugged, as if Sea Eagles in the shower room were an everyday occurrence.

This was not quite what I got from the crew of my RAF Nimrod. At first they naively assumed that the first metre-square cardboard box I loaded on to the plane housed

Sea Eagle nest in a tree, Karlsøyvaer

Sea Eagle nest on cliff ledge, Vettøy

all four eaglets, and were dismayed to discover that there were three more to come. They were quite shocked to see the bird inside standing about half a metre off the ground and weighing 3–4 kilos. Imagine, then, the aircrew's utter panic when I told them that, on this occasion, the Ministry of Defence in Whitehall had declined me permission to accompany them on the flight. The thought of four eagles escaping and rampaging through the aircraft over the middle of the North Sea quite horrified them. Undeterred by the bureaucrats in Whitehall, the RAF crew nonetheless showed me round the plane, mostly to reassure me that they were doing everything to make the eagles' flight as comfortable as possible. As we completed all the paperwork and formalities on the tarmac, the new Air France Concorde landed nearby and I sneaked a photograph, despite the concerns of the young Norwegian security guards.

Having reassured the apprehensive aircrew that the eaglets would lie quietly in their boxes, I watched the Nimrod take off before returning to the staff quarters. My next task was to scrub the overnight production of fishy faeces from the shower walls. It is surprising how high their smelly squirts can reach. Exhausted, I retired to the Mess where I met the camp dentist who spoke English with a Scots accent. He had qualified in Glasgow and practised in Campbeltown for three years before returning home to the military. He took me back to his quarters to gorge on fresh prawns (before discovering that I am allergic to them) and to listen to his Billy Connolly tapes! Not quite the Arctic experience I was expecting.

Within two hours the eaglets had landed in Kinloss, where they were met by Roy Dennis and Martin Ball. It took only another six or seven hours for them to be driven to meet Peter Corkhill, the Chief Warden in Mallaig, with the NCC boat *Rhouma*. Two hour later they had been driven across Rum to their cages.

A Sea Eagle chick in what was probably an old raven nest, being sketched by the famous Norwegian artist Karl Erik Harr, Karlsøyvaer

Sea Eagle chicks in Harald's cellar in Norway, with newspaper on the floor

I, on the other hand took fully two days to make the journey by scheduled flights and public transport. This did give me an opportunity to review my hugely rewarding two-week voyage with Johan. We had landed on 25 islands, visited three communal roosts and 15 active Sea Eagle nests. Eight of these nests contained chicks – four twins and four singletons, twelve eaglets in all. That gave me a figure of 1.5 young per occupied nest. We also heard of a brood of three and were told of a clutch of four eggs. Nine of the sites could be said to be cliff sites (although the only eyrie that could be said to be high and precipitous was the only one not on an offshore island), three of these nests lay virtually on the ground and only one was placed in a low tree, while I heard of another collapsed tree nest in the vicinity.

We recorded 52 food items in or near the nests, made up of 60 per cent fish and 40 per cent birds. 70 per cent of the fish were catfish or what we might call wolf fish (due to the impressive 'teeth'), 12 per cent were lumpsucker, 10 per cent flounder and the remainder various species such

Harald Misund checking a Sea Eagle nest, Nordland

as gadids. Of the bird remains we found, 75 per cent were Eider, the rest in equal proportions being Shags, auks or other species. In the two weeks I clocked up 54 species on my own bird list and could not even begin to count how many individual birds had been Sea Eagles.

I have visited Norway, and especially Bodø, many times now, usually in the company of my close friend Harald Misund. He has showed me scores of Sea Eagle nests within his study area and it quickly became obvious that my fortnight with Johan had given me a slightly one-sided view of Sea Eagles in Nordland. Since this Kommune just north of the Arctic Circle has contributed 140 Sea Eagles to Scotland (Phases I and II), it is useful to have a more accurate picture of the habitat and conditions from which they came. In an attempt to clarify the situation, I asked Harald in 2001 if he would write a short summary of his survey work for the project team's annual newsletter. With forty years of field work behind him, there are few, if any, people who know more than he does about the species in the area. This is what Harald wrote:

Monitoring Sea Eagles in Bodø Kommune, Nordland, Norway 1968–2000

> In the years 1957–1959, the late Dr Johan Fr. Willgohs (Bergen University) found 12 pairs of sea eagles breeding in the outer islands of Bodø (Karlsøyvaer, Bliksvaer and Helligvaer). However, from 1968 to the year 2000, a total of 79 pairs have been located so far, with 40 breeding on the outer islands and 39 on the mainland. Shallow water with rich fishing grounds are important, supporting abundant bird life and a good population of otters. Otters are particularly important to the Sea Eagles, leaving many fish remains for them. When Sea Eagle chicks are small, lumpsuckers are a favourite prey and are easily caught since they are in shallow water spawning at this time.

223 nests of Sea Eagles have been discovered, some of them still in use and known to have been used before the war (1940). Since 1968, 159 of these nests have produced eggs or chicks. Their average height above sea level is 67 m and the average distance from the sea 482 m. Several pairs nest within 500 m of one another, the closest being about 150 m apart. Three pairs have been found nesting within the same square kilometre, while two pairs of Sea Eagles and one pair of Golden Eagles have been found within another square kilometre.

The population of Sea Eagles in Bodø Kommune has obviously been underestimated in the past and, in Willgohs's time, must have been around 60 pairs. Since many people abandoned the outer islands in the later 1950s and 1960s, a further 18 or 20 pairs of Sea Eagles have found suitable nesting territories. Willgohs noted flocks of up to 32 immatures, but in recent times, such flocks number only a handful of birds. In wintertime, however, 15–20 birds normally gather around Bodø harbour and in Saltfjorden.

Some Sea Eagles nest in small trees, such as willow tree (*Salix*), the birds normally seeking food around the islands offshore. Such trees are often broken by heavy snow.

In an area like Bodø, it is difficult to find all pairs and some parts of the countryside are still not well known. It is not possible to find all pairs working in summer alone; it is necessary to work all the year round but especially in February and March. In 120 km from Kunna in Meløy to Engeløya in Steigen, more than 200 nesting pairs have been discovered so far. The location of the Bodø area, which measures 937 sq km, is at 67° 16' N; 14° 24' E.

The earliest eggs in the Bodø area (a clutch of two) were found on 2 February (two eggs in a nest in Karlsøyvaer). Most eggs are laid by 25 March and one even as late as midsummer, when the latest known hatching also took place! Some chicks fledge before mid-summer (23 June) but most leave the nest around 18–20 July.

Chapter 8
Rum

If ornithologists in Britain are ever to enjoy the sight of a pair of nesting sea eagles again it is almost certainly from Norway that the birds will come.
Bannerman 1956

'Would you like to go to Norway for a couple of weeks, and then go to the Hebrides for a month?' I was asked. 'It's to bring back young Sea Eagles for release in Rum.' Well, who wouldn't jump at the chance? Only I could not have known then that the month in Rum would turn into a decade, and Sea Eagles would still remain part of my life thirty five years later.

I suppose the Nature Conservancy Council was uncertain whether this new, 1975 shipment of four young Sea Eagles to the Isle of Rum could be sustained in future years. Perhaps this was why I took only a small rucksack to Rum. Ten years later I had to hire a boat to transport my accumulation of furniture, books, other gear – not to mention a wife and my faithful collie dog Rona – back to the mainland. The lessons from this tale, then, are to take the opportunity if it arises and to be an optimist. This early tentative reintroduction was one of the first. Now, of course, similar projects with threatened birds of prey and other species are commonplace all over the world.

I had first visited Rum on a student field trip, the year after I had met the Sea Eagles on Fair Isle. So I already knew of its disadvantages – only a couple of dozen human inhabitants for company, ferry boats on only four days of the week (weather permitting), at least 250 cm of rain a year and, perhaps worst of all, the biting insects – blood-sucking ticks, ferocious horse flies (clegs) and, last but by no means least, the tiny, persistent and unimaginably numerous midges.

No sooner had I arrived back from Norway to be reunited with my aqualine charges, a press photographer telephoned to ask if he might come over to take some photographs. 'Perhaps I could take a snap of you, with one of the chicks sitting on your finger?' he inquired. I replied that he was quite welcome to come, but had to explain in words of one syllable why I would certainly not risk losing a finger for his photograph. Even the vets who came to examine the birds prior to their release were reluctant to enter a cage with such monsters. Their size was impressive enough but their obstreperous behaviour provoked alarm in the uninitiated. We had carefully avoided the birds becoming used to human presence so that they would adapt to the wild more easily.

Human contact was limited to my short visit each day with food and to check on the birds' welfare. I would spend a lot of time watching from a distance, though, particularly when previous releases were in the vicinity. After five weeks' quarantine, the eaglets were declared by government vets to be fit for release. Such was to be the routine over eleven summers, from 1975 to 1985.

In 1975, it turned out that only one of our first quartet was a male; I called him Odin. He was smaller and more nervous than the three females and also more erratic in his food intake. He began to get weaker and, despite my seeking medical advice over the telephone, Odin died on 8 August. A post mortem by the Government's veterinary laboratory at Lasswade near Edinburgh diagnosed kidney failure but could not detect what had brought about this condition in the first place. I have always suspected lead poisoning, from a bullet fragment hidden in a piece of venison or goat meat. Although I did not feed them anything killed by shotgun and carefully avoided badly shot-up meat, Odin might still have ingested a small piece of lead. On the other hand, the three females – Loki, Karla and Freya – thrived and were released on 26 September, 24 October and 1 November.

For the eleven summers from 1976, and again during Phase II in the mid-1990s, we were fortunate to have the invaluable assistance of Harald Misund. Harald was born in Ålesund just before World War II. His father, after fighting alongside the British in Romsdalen, became involved in organising boats between Britain and Norway (known as the 'Shetland Bus') until he was taken prisoner by the Germans in 1941. Harald remembers watching from his window in the spring of 1945 as an RAF squadron from Scotland bombed Ålesund harbour and thinking to himself that the war must be coming to an end. Unlike so many of his comrades, Harald's father returned home safely after the war but it was his uncle that would take the boy out hunting and fishing, fostering Harald's interest in wildlife. In his spare time, while working at a local fish hatchery, he discovered his first Peregrine nest.

Trained in electronics, Harald joined the Norwegian Air Force in 1955 and with distinctions in all his subsequent training and six months in Denmark, he received his first posting as Sergeant at Gardermoen in Oslo, specialising on the fuel control system of the F-86K. The planes were transferred to Bodø in 1960 where Harald's elder brother lived. Together they would go fishing in the remote mountain valley outside Bodø called Falkeflaug where Marianne, his future wife, lived on a farm. Although it is now abandoned, her family still have summer cabins on that isolated farm, and I have stayed there myself several times. Below the mountain called Falkeflaug (meaning 'where the falcon flies', namely the Gyr Falcon), quite close to the road, is a special valley with lady's-slipper orchids (until recently, our British population had been reduced to a single plant!), Norwegian primroses, twinflower and other botanical gems growing on the limestone outcrops. In addition, *Diapensia* (an Arctic-alpine flower) is abundant on the hill tops; there is only one mountain in Britain where this can be found, near Fort William.

Harald Misund and the author on a collecting trip

Harald began ringing birds in 1961 and spent much of his

Astrid Misund, my colleague Martin Ball and me with our young charges in Harald Misund's garden (Harald Misund)

spare time looking for Sea Eagles and Peregrines from Lofoten, all along Nordland's coast and inland to the Swedish border. He contributed hugely to Projekt Havørn's database in Nordland.

In 1971 Harald was offered a short exchange posting to RAF Leuchars in Fife, where Phase III of the Sea Eagle reintroduction would later be based. From there, trips to Loch Leven, Pitlochry and Inverness, my home town, cemented in Harald a love of Scotland which has never left him. Even without his intimate knowledge of Bodø's Sea Eagles, then, he was an ideal and immediately sympathetic contact for our reintroduction programme.

Surveying potential sites to contribute eaglets for transfer to Scotland is a time-consuming occupation which, leading up to mid-summer and the RAF rendezvous, demanded a lot of commitment. Harald generously gave his time to the reintroduction free of charge, his only remuneration being occasional expenses, donated by the World Wildlife Fund, to enable him to collect from offshore eyries. His wife Marianne, with Astrid, the youngest of their three daughters, always showed interest in the Sea Eagle project, sometimes feeding the captive birds while Harald and I were out collecting more. Indeed, Astrid accompanied her father to Rum in 1985, when they would release the last of the 82 eaglets set free in Scotland, and we would meet up again in 2010 at her home in Florø, she by then a wife and a mother.

The author at Falkeflaug, 1985

There is a much higher incidence of twins amongst Sea Eagles than Golden Eagles

A rare instance of triplets in Bliksvaer, Norway (Harald Misund)

Collecting a Sea Eagle from a nest with twins

As a Captain and Maintenance Controller, Harald finally came to specialise in the navigation systems of F-16s, until he retired from the NAF in 1994. He has, though, continued to support the project and to study his local Sea Eagles. Harald was very influential in promoting Sea Eagles as a tourist attraction in his home town, with sculptures and other initiatives, including the founding of the Bodø Sea Eagle Club. King Harald of Norway is member no. 1, and when invited to give their inaugural lecture in March 1995 I was proud to be appointed no. 23!

The density of nesting pairs located by Harald over the years, and the high incidence of twins, meant that only one eaglet needed to be taken from each nest, leaving the parents with the other to rear to maturity.

In one instance a runt from a brood of three was taken. If left, it would almost certainly have died, but once in captivity it thrived, to be released into the wild once again, albeit on the other side of the Norwegian Sea.

In later years new nests could be selected every time, creating as wide a gene pool as possible. I think we only returned to a nest in a subsequent year on about three occasions.

While they were in captivity, I strived to give the four eaglets as natural a diet as possible, all of it collected locally on the island. Fish, bird and mammal prey were given in equal proportions. During the summer months it was possible to catch an abundance of mackerel offshore and we were sometimes given fish from local fishing boats anchoring in the bay, or from Andy Race at the local fish market in Mallaig. In this way, whiting, haddock, some herring, dogfish and squid were all added to the eagles' menu, with some deep-frozen for later use.

Thanks to their ability to digest most of the smaller fish bones, the eagles derived minerals from their diet, calcium being especially important for the growth of bones and feathers. Fish also contained sufficient moisture to satisfy the eagles' need for water, but the long-term captives had access to water baths in any case. In a climate with more than 250 cm of rain per annum, this seemed an unnecessary luxury!

Apart from occasional small pellets of dried grass, the eaglets regurgitated little waste on their diet of fish. I gave whole bird carcasses as a source of roughage, mostly gulls and crows shot with a rifle to prevent the possibility of the eagles accumulating quantities of potentially poisonous lead shot. Towards the end of the breeding season, young gulls were available from convenient local breeding colonies. With repeated harvests over the years, and as I selected out the conspicuous chicks, the next generation seemed to become more and more subterranean in their habits, hiding in ditches and taking whatever cover they could. My collie dog Rona became a valuable member of the team, adept at sniffing out gull chicks and retrieving eagle prey remains from some plucking ledges that I overlooked or might consider inaccessible. Many of the smaller bird bones could be digested but the larger ones, the feet, beak and feathers were usually picked clean and discarded. In early August scraps of venison, viscera and the heads of shot deer became available from the stalking activities on the island, supplemented by goats shot along the rocky coast.

In 1976 two of the original cages were upgraded and two new improved ones were built along the Harris road. Up to eight eaglets could now be accommodated, while a further four could be tethered nearby. The renowned Scottish climber Hamish MacInnes (in Rum on a filming assignment), gave me valuable technical advice on equptment to use for tethering. Adapting a traditional falconry technique I attached leather jesses to the legs of each bird which met at a metal swivel. From this a length of braided nylon rope served as a leash, terminating in a metal cabarabiner. The carabiner was threaded through a 6 metre length of metal cable, stretched between two short posts. This permitted the eagle to fly from a tree stump perch, just within reach at one end, to a wooden tent-like shelter at the other, whose roof also served as a high perch.

Waste food was removed daily in the interests of hygiene and rats cleaned up any remaining scraps. I never experienced any of the eagles attempting to catch and kill these rats, which

Some young eagles were tethered out in the open, rather than kept in cages

The birds could perch on a shelter at one end or sit inside

The author admiring the new tethering system, Rum

Juvenile Sea Eagle on tethers, Rum

were at any rate mainly nocturnal, but on at least one occasion the tethered male, Ronan, succeeded in capturing and killing an over-confident hooded crow which had rashly strayed within his reach in its search for scraps. At times crows and Ravens would prove a nuisance by pulling food beyond the reach of the tethered eagles. It was my task to make sure that the tethered eagles had generous rations of food so at times I would discourage the corvids with my rifle, meaning that they then ended up on the menu themselves.

Mammal prey was given in the form of deer livers and other scraps from the annual cull undertaken by NCC stalkers; they also gave me an occasional staggie shot because it was in poor condition. I also shot some feral goats on the cliffs and was given a dead calf from the dairy herd kept on the island, or a Highland stirk that had been kicked to death by Callum, the over-zealous stallion with which it shared a paddock. (Indeed, Callum once gave me a nasty and painful bite on my shoulder.)

The amount of food consumed by the eagles varied from day to day, ranging from almost nothing, usually following a hefty meal the previous day, to over 1.4 kg. On average I found that a captive eagle would consume about 10 per cent of its own body weight per day, about 400–500 g wet weight. Two or three day's supply could be stored in the elasticated crop but most of this would pass through the digestive system within a few hours. Once they had fed, the captive eagles tended to sit, lethargically looking around or preening.

Close-up showing the dark head, eye and beak of the juvenile Sea Eagle, Rum

The average daily intake was much the same for the females (571 g net weight) and for both males (456 g) – about 9 per cent of body weight (Table 3). Although food was always provided in excess, the birds themselves were only weighed at the conclusion of the trials so it is not known whether they had increased in weight in the meantime. Being retained in individual cages 4 metres square and 2 metres in height, opportunities for exercise were limited.

Table 3: Daily food intake of captive White-tailed Sea Eagles on Rum (Love 1979)

		Food intake per day				Daily food
Sex	Bird no.	Mean standard deviation (g)	Range (g)	No. of days	Body weight (kg)	Intake as per cent of body weight
Female	1	553.2 ± 272.0	0–1221	106	6.9	8.0
	2	577.2 ± 281.0	156–1221	102	6.4	9.0
	3	584.0 ± 297.0	57–1448	74	5.9	9.9
	4	617.5 ± 191.3	201–1027	20	5.6	11.0
	5	543.5 ± 221.4	261–1031	20	5.5	9.7
Male	6	502.4 ± 178.3	168–820	20	5.0	10.1
	7	408.6 ± 139.2	222–681	20	4.8	8.5
TOTALS						
Females		570.7 ± 290.4	0–1448	322	6.1	9.4
Males		455.5 ± 164.9	168–820	40	4.9	9.3

The results obtained from my Rum study confirmed observations made upon Sea Eagles elsewhere. One study concluded that a fully-grown captive bird, maintaining a constant body weight, required about 500 g of meat per day, about 10 per cent of its body weight. In a wild situation, sustained activity imposes greater food demands, but by how much is not easy to determine. Another study on kestrels indicated that to fly free this small, very active species required two or three times more food than it did in captivity. Willgohs was able to observe that a pair of White-tailed Sea Eagles, together with their three fledglings, each consumed about 625 g of food per day. This was less than one might have predicted, but was considered to have been a slight underestimate since, during the eight-day observation period, all three eaglets lost body weight.

Captive Sea Eagles in Rum could consume two whole mackerel, each weighing about 350 g, within six or seven minutes. Seven or eight smaller fish could be swallowed whole within a couple of minutes. In each instance the crop was filled to the size of a small grapefruit. Meat is more time-consuming to devour, requiring 20 minutes or more before an eagle is satisfied. Willgohs estimated that one wild eagle had eaten two kilos of fish in a single meal while, as we shall see later, a three-year-old, free-flying male on Rum gorged on 1.8 kilos, about 40 per cent of its own body weight (Fiona Guinness, pers. comm.)

The actual weight of prey eaten is, of course, less than that actually killed, due to wastage. Most of the feathers, large bones and often the beak, feet and sternum of bird are discarded, as are the gills, tail and other waste parts of larger fish. Indigestible material such as small bones, feathers or fur which are ingested accidentally are later regurgitated in a pellet. Normally one pellet may be produced every two days or so but the frequency is very much determined by diet. Pellets appear most regularly when an eagle is feeding upon bird prey. These pellets also tend to be the longest and bulkiest (Table 4) since they often contain the head and beak, or a complete leg, wrapped up in miscellaneous bones and feathers. On a diet of small fish, tiny pellets (about 40 x 24 mm) consisting entirely of grass may be produced. Such vegetation, I suspect, may either be ingested deliberately or accidentally while the bird is tearing up prey. The grass may serve to cushion the bird's digestive tract against sharp fish bones. I have even watched captive eaglets tearing at and ingesting newspaper, presumably towards the same end.

Table 4: Dimensions (mm) of 66 pellets produced by three captive female White-tailed Sea Eagles, Isle of Rum 1975 (Love 1983)

<table>
<tr><td colspan="4">Length: 80.7 ± 27.9 (mean and standard deviation) 34–157 (range)</td></tr>
<tr><td colspan="4">Breadth: 31.0 ± 5.6 (mean and standard deviation) 20–45 (range)</td></tr>
<tr><td>Longest: 157 x 27</td><td>Shortest: 34 x 22</td><td>Widest: 114 x 45</td><td>Narrowest: 39 x 20</td></tr>
</table>

In 1976 we were able to approach RAF Kinloss direct, by-passing the cautious defence mandarins in Whitehall, to obtain permission for a civilian to travel back with the eaglets. Peter Corkhill, armed with licences for ten eaglets, flew out to help Harald and to accompany the birds back to Scotland. Martin, Roy and I met them and drove the eaglets from Kinloss to Rum.

The eagles' arrival in 1976 was filmed by ITN for News at Ten. They naively offered their chartered helicopter to transport the ten eaglets from Mallaig to Rum but totally underestimated the task and, in the end, so many TV crew were present that there was only room for one eaglet in its box! The rest followed on by the Mallaig charter boat Western Isles, skippered by Bruce Watt (NCC's own *Rhouma* being temporarily unavailable). Releases began in November 1976 and continued until May 1977, with Isla being the last of the new birds to go, filmed by the BBC for Tony Soper's *Animal Magic*. Sadly, having failed to utilise the food dump, Isla was found dead on the beach on the neighbouring island of Sanday, part of Canna, only a month later. We never released birds quite so late again.

Arrival of the Sea Eagles at Kinloss

A photo call at Kinloss with Martin Ball and the Nimrod crew (RAF photographer)

The author with a new arrival (RAF photographer)

Most of the eaglets arrived in Rum at the age of about eight weeks. They were caged singly or in pairs and only when a bit older could some be tethered out. Unusually, we might receive some younger individuals that demanded hand-feeding, while one or two of the most reluctant older birds needed to be hand-fed initially.

As soon as the eaglets could use their feet for moving around and tearing up prey, they would threaten any human approaching. Once they could fly – at ten or eleven weeks – they became distinctly nervous at my approach. We did not want them tame or imprinted upon humans.

Immediately prior to being given its freedom, each eagle was caught and hooded with a specially-made falconer's hood, which subdued the bird as it was being handled. Measurements were taken of bill length from the cere to the hook tip and of

A newly-arrived chick, younger than the others, Rum

Installing the birds in their cage

The author inspects a new arrival, Colla, in Rum

the bill depth from the cere to a point perpendicularly below the lower mandible as it is held closed. From 1977 additional statistics were gathered on the thickness of the tarsus at its narrowest point midway along its length. Adequate samples of wing, tail and tarsus length can be found in the literature and in any case abrasion and age can confound the measurement of feathers. Finally the bird was wrapped in a canvas sling for weighing, allowance being made, of course, for the sling and for any food that might be detectable in the crop. In all but three instances the crop was judged to be empty.

Together these five statistics (bill length and depth, tarsus width and thickness, and body weight) were usually sufficient to sex the eagles and in most cases were found to confirm my own, purely subjective, assessment based upon the bird's appearance and

Hand-feeding an unusually young Sea Eagle chick

As soon as they were able to use their feet they could feed for themselves

1975 female Karla in threat pose before release, Rum

behaviour. Several birds fell within a zone of overlap between males and females but I consider in retrospect to have sexed only three birds wrongly, early in the project when I was less experienced and had less data to hand.

Each bird was provided with a British Trust for Ornithology (BTO) numbered metal ring on one leg so that, if found, the origins of the bird could be traced. On its other leg I provided each eagle with two coloured plastic Darvic rings, sealed with chemical fixative. These facilitated individual recognition in the field. The colour of the lower ring denoted the year of fledging and the top colour made the combination unique to any one individual. In 1976 I used a variety of combinations but placed them all on the right leg; in all subsequent years the colour

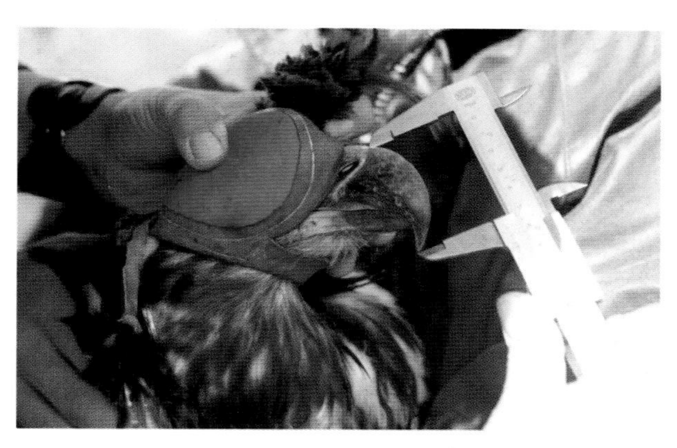

A falconer's hood made handling the birds easy during procedures such as measuring

A BTO Ring is applied to one leg of each chick prior to release but, nowadays, a European colour ring has replaced a plastic colour ring combination on the other

rings were placed on the left leg, with the BTO ring on the right. From 1982, a coloured and numbered tag was attached to each wing, the colour again denoting the year of release with the large single number (0–9) applied to one individual. In this way, individual birds could be more easily distinguished in the field and their fortunes more closely followed. The materials had been made available by a university friend of mine, Dr Mick Marquiss, by then a colleague of Ian Newton's at the Institute of Terrestrial Ecology (ITE).

With the arrival of ten eaglets in June 1976, the project team was considering retaining on tethers four of them, two of each sex, with a view to their later breeding in captivity in a large cage facility to be built on Rum. We would in that way become less dependent upon the supply of wild birds from Norway. The 1976 females Colla and Sula, together with a male, Ronan, were to be retained in long-term captivity with a fourth bird, Beccan, which I wrongly judged at the time to be a male. Over the succeeding months I had to refine and strengthen the tethering furniture (to use the falconry term) because Colla and then Beccan made bids for freedom. Quickly, using bait, I managed to lure Colla close enough to catch her in a long-handled net. Next, some weeks later and after several days on the run, Beccan was caught on the raised beach at Harris, after a colleague, John Bacon and I – armed with a powerful torch and net – managed to dazzle and trap her. She went on to make yet another escape, though, and this time, at mid-summer without the cover of darkness, could not be recaptured. Her place was taken by the male Cathal but Beccan continued to serve a useful function in mentoring the 1978 releases at the food dump. This doubtless enhanced their survival and, as sea eagles are such sociable creatures, underlined the importance of continuing to release more birds in the future to build up a population of mixed ages and experience.

Patagial tags were fixed on each wing, the colour signifying the year of release while the number, letter or pattern denote the individual

The team was host to Dr Willgohs in early June 1978 and he was able to visit me in Rum to view the four 1976 eagles still captive, along with some of the released birds. The RAF took Morton Boyd and Roy Dennis to Bodø in June, allowing them a couple of days to meet with Harald and the various authorities locally who were involved in the project. Meanwhile, basing their own ground crew in Bodø, the Nimrod carried out some surveillance off the North Cape before returning with Roy, Morton and eight eaglets on 21 June.

Soon, however I noticed that one of them, a female I named Shona, seemed to be suffering from a mineral deficiency. A vigorous bout of exercise not long after her arrival seemed to lead to the fracture of both her legs. A visiting doctor temporarily set her brittle bones until the eagle reached a vet on the mainland, and eventually the Veterinary College in Glasgow, where the fractures were pinned surgically. Carol Scott and George Watt, two local birders who maintained a rescue centre (appropriately enough in Eaglesham) let Shona convalesce and under their expert care she was returned to Rum in October 1978. By then, however, she had lost all fear of humans and still walked awkwardly. She had strong callouses around the fracture sites but then developed problems with her wing joints. Incapable of normal flight, she finally caught a chill, perhaps pneumonia, in September 1979 and died. This time a prompt post mortem at the North of Scotland College of Agriculture in Inverness revealed a severely congested bile duct and gallstones which could have contributed to, if not caused, her death.

Tragically, after a period of exceptionally wet and stormy weather, the beautiful Colla, our largest ever female and one of my favourites, suddenly took ill in September 1978 and died a few days later. Not only were we unable to summon veterinary help in time, but we were prevented by further transport delays from despatching the corpse for post mortem and pathological examination. We suspected that Colla had died of a bacterial or viral infection, perhaps transmitted by the rats or crows which frequented the tethered site for scraps.

We were unable to release the injured Shona but the remaining seven birds were set free on 15 and 16 October and 15 and 24 November. The presence of Beccan and the occasional previous releases flying free, along with the four tethered birds from 1976, encouraged many of the 1978 birds to remain together and freely utilise the food dumps.

However, the sad losses of Odin, Colla and Shona highlighted the remoteness of Rum from speedy and specialised veterinary advice, deficiencies which could also prove crucial should we set up a captive-breeding facility on the island. At that time, too, financial restraints were imposed on NCC as a government-aided department, and we were conscious that Harald wished all the birds to be liberated. He assured us of further young in future years so we decided to abandon the captive-breeding proposal altogether. In the spring of 1979 Cathal, together with the original captive pair from 1976 – Sula and Ronan – were finally set free. Sadly, Cathal was to be found long dead in August 1979 on exactly the same Sanday shore where Isla had been discovered three years before, no doubt a quirk of tides and currents. The cause of his death could not be determined but he was another bird who rarely utilised the food dumps. His was the third recorded death from among 23 birds released so far.

I travelled to Bodø with the RAF in June 1979 and returned

Colla, our biggest and most beautiful female, Rum

with six more Sea Eaglets. Conscious now of the role of sociability for the species, I was determined to release them as early and as close together as possible, but the vet who first had to clear them was delayed for three weeks by bad weather. The first two were set free on 24 August and the remainder during September. As I expected, they were often seen in the company of some older birds around Rum but, without the presence of poor Shona on tethers, sightings were now also becoming more common elsewhere.

I nonetheless continued to enjoy some fascinating interactions in 'my' local flock. On 20 June 1980, Martin Ball and I travelled with the RAF to Bodø, where we were involved in filming with the BBC Natural History Unit. With his own work pressures, Harald in particular found the process a bit frustrating, but we managed to collect a further eight eaglets for Rum, including a runt from a brood of three.

The first two Sea Eagles were released on 27 August, with Hugh Miles from the BBC arriving to film further releases during September, with the last freed on 2 October. By this time the new releases had attracted some of the older birds back to the cages, including my old friend Sula from 1976, now approaching adulthood. She too was captured on camera and the resulting programme, narrated by David Attenborough, featured on BBC TV's *Wildlife on One* series on 8 September 1981 and was well received. One particular release sequence still regularly features on television and, given the now antiquated design of Hugh's slow motion camera, I still don't really know how he managed to film it. He had already set a regular camera running on a tripod behind me, while he stood in front with his unwieldy slo-mo to capture the whole take-off from the moment I whipped the hood off the bird's head.

With NCC suffering dire cutbacks in 1980, a grant from the World Wildlife Fund was being administered through the Scottish Wildlife Trust, with my friend and colleague Jeff Watson working as their Development Officer; for that year I was officially employed by them. And only now, after Roy and I had shown their Scottish Director Frank Hamilton free-flying Sea Eagles on both Rum and Canna, was Roy finally able to represent the RSPB on the project team. It was Beccan the escapee, who had been living all this time on Canna, initially still with a tether dangling from one leg and by now a full adult with brilliant white tail, who swung it in the end.

Hugh Miles, filmed Sea Eagles for BBC Wildlife on One *in 1980*

In March 1981 my new contract, still short-term, was announced, while the RSPB had won funding from the Eagle Star Insurance Group to employ a temporary warden, Colin Crooke, to begin surveying Sea Eagles in the wild. Courtship, stick- carrying and even copulation were now being observed and two pairs of 1976 releases had established territories. One setback, though, was the discovery in April 1981 of Erin, a male released the previous year, poisoned in Caithness, the fifth death among 37 birds released into the wild. Analysis of both bird and bait revealed Phosdrin; Erin was otherwise in good condition. In October of that year, another poison casualty was to be found dead in Skye; this was 47 km north of Rum, where it had been released exactly four years earlier. This bird was a particularly sad loss as it was almost sexually mature. On the plus side, more extensive surveys away from Rum were turning up at least 25 distinct individuals still alive, giving us a very encouraging minimum survival rate of 66 per cent.

Roy Dennis and Frank Hamilton, RSPB's Scottish Director on Canna, 1980

A poor breeding season in Norway meant that it was not possible to take as many eaglets as our 1981 licence allowed. I went to Norway with Roger Broad, Roy's new RSPB assistant whom I had met when he succeeded Roy as Warden on Fair Isle in 1970. We returned with only five eaglets to take to Rum. They were all successfully reared and released in late September and early October, in the presence of up to three older birds.

In 1982 we had yet to detect any breeding attempts in the wild and another 1980 release, a female called Hynba, was found dead on Canna in May. In view of our successes to date, Harald Misund and the Norwegian authorities agreed to step up efforts for a final push by supplying ten eaglets annually. For the first time, though, because of serious commitments in the Falklands following the invasion by Argentina, the RAF was unable to assist us. Fortunately, an Oriel aircraft from the Norwegian Air Force stepped in and transported ten eaglets to Kinloss, accompanied by Captain Harald Misund in a more official capacity!

Again due to bad weather, veterinary clearance for the 35-day statutory quarantine could not be obtained until mid-September. By this time I had got married (it lasted ten years) and my family had arrived on the island for the wedding. I took my young niece Heather to see the tethered eagles and her other uncle, my elder brother Jim, drew an amusing cartoon of the occasion!

A cartoon by the author's elder brother, the late Jim Love, commemorating his niece seeing the eagles in Rum, July 1982

I had been able to address a major Bald Eagle conference in Rochester, New York State, and build this into our honeymoon. There I met for the first time a host of delegates, mostly American, who were involved in a similar conservation project with our eagle's sister species. On Sunday, 15 August we had an early start for the Bald Eagle hacking station at Oak Orchard. There we were lucky to see one of the previous year's released birds which had been in the area for several days. Pete Nye, the project director, gave us a close-up look at the hacking tower, with 20 young Bald Eagles from Alaska in the cages awaiting release. The cages were on stilts overlooking the wetlands, the captives constantly monitored by CCTV cameras from the comfort of portacabins a short distance away. The birds looked remarkably similar to our own Sea Eagles of that age. This project had begun in 1976, a year after our own project had started, but this was technically restocking the state, since the Bald Eagle still occurred in many other parts of the USA (Nye 1982).

Eight of the ten 1982 Sea Eagles were released carrying patagial wing tags, which greatly facilitated individual recognition in the wild and revealed some interesting movements which had previously gone undetected. The male Gregor, for instance, released on 1 October, remained in Rum for nine days before being seen on the neighbouring island of Muck on 13 October. He eventually turned up near Stornoway on the Isle of Lewis on 19 October, no doubt aided by southeasterly gales. He then remained on Lewis until the following spring. Tagging also revealed just how many different Sea Eagles were returning to utilise the food dump, a feature that had been underestimated when identification relied solely upon the less visible colour leg rings.

Through the good efforts of my new boss Derek Langslow, the successor to Derek Ratcliffe as Chief Scientist, my own contract with NCC had been assured for a further three years, although I had first to apply for my own job! Eagle Star Insurance agreed a further year's funding for surveys by the RSPB.

Early in 1983 there were several sightings of young Sea Eagles in southern England but these were genuine immigrants from the Baltic. Back in the Hebrides, Richard Coomber, an ornithologist in Mull, had been reporting some of our released birds and regular pre-breeding activity for several years. At last two nests were located during April and subsequent breeding activity will be described in the next chapter. As promised, ten eaglets were again despatched from Norway in June when Roy Dennis and I were able to fly once more in the Nimrod. In the meantime, Sula had paired up on a territory along Rum's southwest cliffs but as soon as the new arrivals were installed she appeared back at the cages, once bringing a dead Fulmar for the eaglets inside. She was unable to get it through the mesh and finally ate it

With Pete Nye, Director of the New York State Bald Eagle Restocking Project, 1982

Alaskan Bald Eagles awaiting release in New York cages, 1982

herself. All ten imports were ringed, tagged and released during August. Most returned to the food dumps for several months, except a male, Fergus, which was found dead in the north of Rum on 21 October.

At the end of the season I moved from Rum to live near Inverness but returned to check up on the situation on the island at regular intervals. Roger Broad, meantime, had become the RSPB's Regional Officer for Strathclyde and, given the regularity of records from the Argyll islands in his patch and the imminent likelihood of Sea Eagles breeding there, he was invited to join the project team. Roy and I had continually liaised with the media on the project's behalf while I regularly submitted articles to magazines and journals. And, of course, we now had the *Wildlife on One* programme on release, with fresh national news and TV coverage whenever new birds were imported into Kinloss. The RAF crews always enjoyed the publicity too, with Flight Lieutenant Steve Rooke playng a pivotal role in the continued good relations with the project. My book *The Return of the Sea Eagle* was published by Cambridge University Press later in 1983, slightly prematurely on the breeding front, as it happened, and with a couple more years of imports yet to be made.

Tower block of release cages for Bald Eagles in New York State, 1982

Our birds were now making their first breeding attempts. After several prolonged visits to Rum and Canna over the winter to monitor the fortunes of the previous releases, I returned briefly in June to prepare the cages for the latest arrivals. Martin Ball and I travelled to Norway with the RAF, devoting some considerable time to the Norwegian press, two of whom came to Rum later in the summer to follow up their story. All ten eaglets settled in well and were filmed by our own TV news crew on 14 August which, together with an interview with me, featured on both ITN and Channel 4 on 20 August.

I released six eagles together in early August 1984 and they remained in a flock, even without the appearance of Sula and her mate who were busy on territory elsewhere on the island. The remaining four juvenile Sea Eagles were set free on

Sula on cage, July 1983

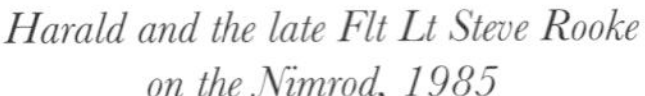

Harald and the late Flt Lt Steve Rooke on the Nimrod, 1985

Johan Willgohs and the author in Mallaig harbour, 1985

14 and 16 August and by the end of October, most of the year's intake were independent of the food dump and beginning to move further afield. I returned to Rum as often as I could but, living on the mainland, was well placed to assist occasional survey work elsewhere in the Hebrides. Eagle Star again came up with the cash to continue monitoring by the RSPB, while Britoil financed a colour brochure which I wrote and which NCC published. Meanwhile, Jeff Watson, now about to join NCC himself to study Golden Eagles, was undertaking a useful analysis of Sea Eagle diet. He would of course eventually write the definitive monograph on the Golden Eagle (1997) but tragically died of cancer in 2007, having almost completed the second edition (Watson 2010).

In 1985 Colin Crooke (now Roy's new assistant at the RSPB Highland Office) and I returned from Norway with the final batch of ten eaglets, all of which fledged successfully. Johan Willgohs and his wife Anne-Marie visited us in Scotland.

Later in the year Harald Misund was able to visit Rum for the release of the first six eaglets:

> Now, on a fine August morning, my youngest daughter Astrid and I are in John Love's Land Rover, on the island of Rum. Releasing the young eagles today will mark the end of the Reintroduction Project. We stop near some Highland cattle grazing and immediately an immature Sea Eagle appears overhead. I have a warm feeling, and tears come to me easily. So they are back. I have sent over 80 eaglets to Scotland, but always, when stowing each bird in its cardboard box for the flight to Scotland. I have felt a little sad to feel its heart thumping, its brown eyes almost bewildered and fearful. Often, I asked my family at home if I was doing the right thing. So releasing this last bird lets a burden fall from my shoulders. It is also very exciting.

Harald Misund releasing a Sea Eagle in Rum, August 1985

The remaining Sea Eagles were set free on 16 and 18 August during an extremely wet summer. Two Norwegian journalists visited Rum a few days later to report on the eagles, so it was fitting that this was the year that the first Gaelic-speaking Sea Eagle fledged in Scotland for over 70 years. Over eleven seasons, a total of 85 eaglets – 40 males and 45 females – had been imported from Norway during Phase I on Rum. Three eaglets (5 per cent) – two females and a male – died in captivity. One might have expected a small proportion of 'non-viable' birds which would be eliminated from a wild population during fledging or immediately on leaving the nest. In captive conditions such lower quality individuals may survive longer but would still be likely to succumb on release, so we could not be too discouraged that some losses occurred at this stage of the operation. We now had our first breeding pairs and expected our wild population to slowly build up numbers. That story will be told in the next chapter.

Chapter 9
Adapting to the wild

Is mairg a théid do'n traigh 's na h-eoin fhéin 'ga trèigsinn.
It is a pity for the one who goes to the shore when the very birds are deserting it.
Gaelic proverb

The instant I slip the leather hood from its head the eagle blinks and spins round to threaten my movement behind. One eye stares at me suspiciously while the other slowly begins to register unfamiliar surroundings beyond. At last the eagle opens its wings and with a loud 'swish' is borne away with the wind. At this moment in each release I am conscious of the eagle's seeming to hesitate, expecting itself to be tugged back to earth by its jesses…but no, this time it is free. It beats its ponderous wings to gain height, seeking control of the wind with its straining pinions; its legs dangle awkwardly beneath for never before has it had occasion to tuck them up. On this, its first sustained flight, it spontaneously discovers the rudimentary principles of flying. When its undercarriage is finally stowed the bird is more stable against the sky and negotiates a huge semi-circle. A passing Raven stoops idly from above but the eagle can counter defiantly by tilting on one wing to flaunt its outstretched talons. With a few more flaps the eagle circles again, gradually loses height, wobbles uncertainly and tumbles into the long grass. It had chosen a gentle, thickly vegetated slope free from rocks and other impediments for this, its first landing at high speed. Hastily the grounded bird recovers its dignity and, sitting upright to survey the landscape, it seems to register distinct bewilderment at this strange new turn of events. Thirty minutes later it begins to shuffle laboriously through the long grass to reach an inviting knoll nearby, where it then remains for several hours.

A Hooded Crow appears beside the eagle, expectant of a tasty scrap, but perceiving no food it boldly delivers a hasty tug of annoyance at the huge bird's tail before fleeing. The eagle reacts sluggishly and, swivelling round, it catches a gust of wind under its half-opened wings and is lifted off on its next brief flight. The sorties become longer and more adventurous as the eagle makes its way along the hillside. Finally it is dark and the bird displays little initiative in its choice of roost, preferring simply to move into the lee of a large boulder where it spends the night.

At times, however, a release proved an embarrassing anti-climax. One eagle chose to walk out through the open door of its cage and set off, with a hunched but determined shuffle, to scale a prominent knoll nearby. If possible, I preferred to liberate eagles when birds from previous years were in the vicinity. This often stimulated more excitement so that on a maiden flight a youngster could be induced, as it normally would be by its own parents, to execute quite complex manoeuvres. If sometimes lacking in drama, though, the moment of release is thrilling, certainly for me and, I can't help but feel, for the bird too. Such magnificent creatures were not brought into this world to be held captive; it was especially rewarding to accord them their freedom, knowing that one day some might choose to breed on our long- forsaken shores.

The author releasing a young Sea Eagle, Rum

A young Sea Eagle, newly liberated, Rum

Although the eagles may have been able to fly within a few weeks of arriving in this country, they were retained in Rum for a further six to eight weeks. Part of this period was the mandatory month of quarantine and during the rest we might have expected them to be most dependent, both physically and psychologically, upon their parents. During the project only eight were released during the spring months, four of them being the long-term captives Ronan, Cathal, Beccan and Sula. Cathal did not survive.

Food was always available to them in the vicinity of the cages or tethered site, of course. The presence of eagles which were still captive usually attracted the youngsters to return. Their social nature was evident, these now-wild Sea Eagles often opting to steal food from the tethered birds

A young Sea Eagle flying over the cage, Rum

rather than to feed on their own at a deer carcass lying only a short distance away. Once a routine was established the 'decoys' themselves were liberated. The juveniles usually teamed up together and often, too, with older birds released in previous years, benefitting from their experience. It was reassuring to see how compliant the old birds were, some even willing to yield up food to the juveniles. Their survival prospects were thus much enhanced, an important by-product of the project's continuaton over several years.

The very first Sea Eagle released in Rum had no such advantage. I had called it Loki, suspecting it to be a male, but once fledged the bird was obviously a female. On 27 September 1975 I opened up the door of Loki's cage but, frustratingly, found her reluctant to emerge. A fresh attempt was made the next morning by removing the wire from one wall of her cage. We watched from afar as she leapt at where the wire had been and literally fell out of her cage to freedom. Frantically flapping her wings she became airborne and found herself carried across the glen, dangling her legs absent-mindedly, before crash-landing on a far slope. It wasn't quite the drama we had anticipated for our first release. After several minutes she regurgitated a fat pellet. I remembered how she had gorged on two Red Grouse the day before, which would explain her lack of co-operation that day. Before dusk she made only one or two brief exploratory flights but the next morning I spotted her soaring competently at the top of the glen. Twice she was attacked by a juvenile Golden Eagle but skilfully avoided it in the air. Ten minutes later two adult Goldies joined in the fray but soon lost interest and moved on.

One of the new intake of chicks from Norway on tethers, Rum

Within two days I found a Great black-backed Gull carcass had been carried from the food dump and on the fourth day I watched her eat some deer liver that I had left for her. After that she began to wander further afield but would regularly return to sit on top of the cage which still contained the female Karla.

By mid-October Loki had discovered Freya's cage but would not have realised that she was the female with whom she had shared her nest in Norway. She readily accepted food left at the cages but once Karla and Freya were set free – on 24 October and 1 November respectively – neither cage held much attraction for her and I then established an additional food dump nearer the shore. By now all three eagles were finding deer gralloch (the discarded entrails of shot deer) which they readily located on the open hill. During November Karla and Freya were seen together but after 31 October Loki must have left the island; her body was found on about 19 November beneath power cables in Morvern, Argyll, some 64 km to the south of Rum. Unable to recover the carcass we could never positively ascertain the cause of death. Perhaps the power lines were a cover to hide a more dastardly crime – poison, perhaps – although later we learnt that a local suspected she had been shot.

The other two females remained on the island until the following February and one was still present two months later. On 3 August 1976 I received a reliable report of a Sea Eagle at Arisaig, on the nearby mainland. One returned to Rum later that December and was seen on neighbouring islands in April and May 1977. It was possibly a different bird – by that time in third-year plumage – that was sighted on Islay far to the south on 24 February 1978 and three months later on the Antrim coast, just across the water.

In June 1976 we had received a license to import ten eaglets from Norway and the first pair were released on 23 November. Ailsa, whom I now believe to have been a male, returned to the food dump within two days but another male, Fionn, was seen only rarely. Both were seen feeding at a deer carcass at the end of January 1977 and were still in the vicinity during March. A male, Cuillin, was freed on 24 January 1977 and another, Brendan, on 28 February. Both were observed to use the food dump for two or three weeks. A female, Isla, liberated on 15 April, was at once chased from the area by persistent mobbing hooded crows and again, later that same day, by inadvertent human disturbance. The weather during the ensuing week was wet and misty so that Isla was unable to find her way back. I suspect she may have starved, for exactly a month later her decomposing remains were washed ashore on the Isle of Sanday, at Canna, 12 km to the west.

As soon as Iona had been released from her cage on 3 May 1977 it had begun to rain. This prompted her to leap into a puddle and lie first on one side and then the other as she splashed herself with her wings. She soon became so waterlogged that she was unable to take off and had to seek temporary shelter to dry off behind a large rock. For the next couple of weeks she frequented the shore and by 16 May had moved a few kilometres around the corner to Papadil, at the southeast of Rum. Here, on 22 May, I saw her indulge in yet another bath, in the shallows of a freshwater loch. It is interesting that in later years I discovered eagle footprints in the sand along the shores of other lochs inland, suggesting that bathing may be a favourite occupation of certain individuals.

After the following spring, sightings of birds released in 1976 became less frequent, although two were seen on the mainland nearby. On 7 August 1977 one was observed at sea between Skye and Rum; yachtsmen saw it land momentarily on the water amid a flock of auks, possible attracted by a shoal of fish.

The four eagles which were retained as potential captive breeding stock were tethered within sight of one another in a fenced enclosure, to keep them away from curious cows, deer and people, on an open hillside near the cages. This site was to become a focal point of activity in subsequent years. In March 1977 Brendan began to steal food from the tethered birds so a food dump was established there.

Tethering allowed the birds to fly along a stout wire, between the shelter and a perch

Female Beccan during her first escape, Rum

Because the tethered birds were never allowed to become tame they exerted considerable pressure on the materials restraining them. The first escape was achieved by Colla, who broke the metal ring at the end of her leash. She flew competently from the outset and by dusk had made her way to the shore, choosing to roost on a low, accessible rock. I memorised the exact location and the easiest approach route before going to summon help. Armed with powerful torches and long-handled nets, Peter Corkhill and I set off silently in pitch darkness. We arrived at the appropriate rock where, at the last possible moment, I switched on the torch to find Colla dazzled by its beam. She took off just out of reach of our nets, but fortunately became confused in the beam and came back to earth again only a few metres ahead. Immediately she was in Peter's net and within minutes was safely secured on a reinforced tethering wire.

Beccan escaped next, on 13 October 1976, circled competently several times and landed on a distant cliff face. She took off again and we soon lost her in the rough terrain. It was three days before she turned up again when, as with Colla, I was able to follow her to roost. Repeating the previous escapade, John Bacon and I approached within reach and dazzled Beccan in the beam. She took off, her weight breaking the handle of my net, but she was tangled in the mesh long enough for me to lay hold of her, so once more a fugitive was returned to captivity. On 6 December I noticed that Beccan had pecked part way through her leather jesses and I had to renew them.

Her ultimate triumph came on 4 May 1977 when, at a year old, she again escaped, evading all subsequent attempts to recapture her. The dazzling technique proved ineffective because the summer nights were bright enough for her to detect our approach. This time she chose to roost in steep, rocky terrain which was difficult to cross quietly under cover of darkness. Consequently I changed tack and laid out some bait on which were concealed nooses of nylon to entangle the eagle. Beccan was not to be deceived easily, however, and merely helped herself to food from one of her still-tethered companions. Somewhat embarrassed, I temporarily removed all other food so that this time Beccan had to feed at the bait. Once satisfied, she tried to take off but found herself held. I sped to the scene and had almost reached her when she broke free. Two nooses had been pulled taut but over her claws only, so that the nylon merely slipped off when she struggled. During June, Beccan wandered further afield and rarely returned to the tethered site thereafter. The leash dangling from her legs proved no hindrance and within a month Beccan was regularly catching live prey. Her escape was to prove fortuitous. Jesses and leashes formed a conspicuous marker so I received many reports to indicate her movements. She visited the neighbouring Small Isles and Ardnamurchan, on the nearby mainland, often returning to Rum to provide fascinating interactions with other Sea Eagles, as we shall presently see. Within fourteen months only a few centimetres of leather remained around one leg.

In June 1977, with three eagles yet on tethers, a further four were imported from Norway. The first pair – Mungo and Vaila – were released on 18 October; they soared around together and suffered a brief attack from a passing Golden Eagle. They displayed a propensity to wander and only Vaila would return occasionally to utilise food dumps. Later that month Mick Marquiss arrived to demonstrate radio-telemetry equipment. Although state-of-the-art at the time, this was pretty primitive compared with today's technology. We made up a small 20-gram transmitter encased in epoxy resin. This radio was tied at the base of one of the female Gigha's central tail feathers. To prevent the 12-inch wire aerial whipping around and possibly snapping, it was threaded through part of the central shaft of the feather. Once the tail coverts were smoothed back into place the radio was invisible, and over the next few days we ascertained that it did not trouble the bird, nor had she attempted to remove it. Since Gigha herself weighed 6.1 kg, the additional weight of the radio was insignificant..

Radio transmitter being fitted to Gigha's tail feather, Rum

On 4 November 1977 Gigha was given her freedom. After several brief flights she landed beyond view, but I could pick up a signal using the portable receiving equipment and by taking a cross-bearing I could plot her approximate position. She did not move again before dusk and was at exactly the same spot the next morning. The heavy rain and mist inhibited my finding her with binoculars, although I could detect her brief sorties by the fading and surging of the signal; once she circled for about 15 minutes. The day remained wet and misty and she made only one or two short flights. During one which lasted 35 minutes she was attacked by two Golden Eagles but countered in the usual fashion by swerving to one side to present her talons. After another flight of 20 minutes, she disappeared over the crest of a distant hill. It was two days before I located her again, in a remote glen on the east of the island, but thereafter her movements became difficult to follow since she frequented high ground. On 16 November I had to terminate the trial but Gigha remained on the island for the rest of that winter.

Young Sea Eagle on being released, Rum

In direct line of sight I had found that I could detect a signal from a distance of 5–6.5 km, but on an island as large and hilly as Rum, such opportunities were rare. The best signals were received when I positioned myself on one of the high hills but in so doing I sacrificed the mobility needed to obtain useful cross-bearings or to move position to follow the bird's signal. Often the smallest features could totally mask the signal. During the first four days that the eagle had been under continuous surveillance – 2,080 minutes in total – Gigha made only 21 flights totalling 172 minutes, so only 8.2 per cent of her time had been spent in the air.

A second brief trial was attempted in 1978 with two of the eight eaglets imported that year. A female, Risga, and a male, Kieran, were both fitted with tail-mounted radios and were released on 13 and 19 September respectively. They were tracked in the vicinity of the food dump until the end of the month when both radios suddenly and simultaneously ceased to function. That month the rainfall was 66 per cent more than average and we suspected that dampness had proved too much for the waterproofing of the radios. On 29 September, 16 days after her release, I found Risga on the enclosure beside the tethered birds, her plumage sodden. This hampered her clearing the surrounding deer fence and allowed me to lay hold of her. She was in good health but I kept her in captivity for several days and, after some good meals, released her again on 2 October.

Two males, Fingal and Conon, were set free on 15 and 16 October 1978, followed by three females – Ulva and Danna on 15 November and Aida on 24 November. Morton Boyd was present for these releases, as he would later recount in his autobiography. At this time Beccan reappeared in Rum and came regularly to feed alongside the captive birds. It was often difficult to identify the various juveniles using the coloured rings but their plumage also proved useful. Several birds had damaged or missing flight feathers, were in moult or else showed distinctively pale or dark plumage.

Beccan on tethers in her first year, with black beak, dark eye and chocolate brown plumage, Rum 1976

Beccan in her third year with lighter beak and mottled plumage, Rum 1978

Two years older than the others, Beccan was conspicuous due to the contrast between her pale, worn body plumage and fresh, newly-moulted feathers, which gave her a distinctive mottled appearance. Her beak was quite yellow, as was that of the similarly-aged Ronan, compared with his captive sibling Sula; Ronan's voice also became more shrill in pitch about this time. Over the ensuing year, as Sula's beak became more yellow, she retained a noticeable dark spot on each side; this was still detectable as a deep, brownish-yellow patch even in her fifth year, greatly facilitating her identification in the field.

On 24 November 1978, immediately following Aida's release, a distinctive eagle appeared on the scene. She had markedly pale plumage, almost white on her back and wing coverts. Several secondaries on each wing protruded an inch or two beyond the others and would seem to have been retained from her first year plumage. I eventually managed to distinguish her coloured rings which identified her as the female Gigha, released in 1977. (I sometimes wonder if Gigha might turn out to be the pale female that Dave Sexton in Mull came to call Blondie, although he thought her to be a couple of years older). Over the next month or two it was not unusual to see four or five eagles in the air together, or coming in to feed alongside the tethered birds. Several roosted on a large rocky knoll near the food dump. Beccan was a central figure in this aggregation; she often flew wing tip to wing tip with Gigha, several times swooping to present talons, once turning a complete somersault like a raven but with her wings still fully extended.

On 4 December 1978 I watched Beccan attempt to take food from the tethered female Sula, who defended herself vigorously. Instead, Beccan approached the male Cathal, who submissively permitted her to feed. As soon as Beccan took off with some food, she was pursued by the juvenile male Fingal, who screamed hungrily. Both landed on a nearby slope but Beccan took off again at once, leaving the meat for Fingal; she procured herself another portion from the food dump.

The tether site in winter, overlooking the Rum Cuillin

Second year female Sula at her tether, Rum

The following day Fingal was again seen to pursue Becccan, screaming and swooping up from below in an attempt to snatch the prey from her talons. Eventually he succeeded but as soon as he landed to feed, he was dispossessed by three other youngsters, one of which consumed the food.

Such interactions became more frequent but the eagles were showing less attachment to the site and were wandering to other parts of the island. In early January 1979, during a period of exceptional frost and snow, Beccan and at least three other eagles turned up on Canna where they were seen to feed off a sheep carcass. Gigha and Ulva remained in Rum to be rejoined two weeks later by Beccan and the obsequious Fingal. I once watched Sula leap angrily at Beccan to keep her at bay, but after two or three such attacks Sula had moved sufficiently far along her tether to allow Beccan to pounce on her food unmolested. Beccan gained height with her ill-gotten gains, with Ulva in tow, begging for food and getting some reward.

On 6 February both Cathal and then Sula on their tethers successfully fended off Gigha, but a wily Hooded Crow did manage to make off with a small fish. Gigha immediately gave chase so the Hoodie had to drop its loot for fear of its life. Gigha retrieved the fish and rose to 100 m or so to hover into the wind and consume her prize directly from her talons. In this way she evaded a half-hearted attempt by the juvenile Kieran to snatch it from her.

The next day Fingal was also seen to retrieve a scrap dropped by a Hooded Crow but Conon had to challenge both Sula and Cathal before he finally won some fish, which he hastily snatched in his beak; he took to the air and deftly transferred his prize to his talons. On 19 February, while I was replenishing food supplies at the tethered birds, Conon appeared above, holding a clump of grass in his talons. He flew around clutching this for some time, perhaps in an act of impatience and frustration at my presence which was denying him access to the food dump. I retired to watch him swoop down on Ronan, hardly pausing as he grasped a talonful of small fish. At the same time he had ripped up a quantity of dried grass, so hovered into the wind to pick out and swallow some of the fish. This he achieved with less of the accomplishment shown by Gigha the day before, at a lower height and dropping several scraps in the process. When he swooped low for more, he induced much excited calling from Sula on her perch.

It was encouraging to see the continuing presence of three or four of the seven eagles released five months previously, the period of freedom which would doubtless be the most hazardous in their young lives. They undoubtedly benefitted from the example set by older birds such as Beccan and Gigha, who lent a certain cohesion to the group. The continued attraction of the food dumps was further enhanced by the three eagles – Ronan, Sula and Cathal – still tethered nearby.

Second year male Ronan, Rum

By this time, however, we were conscious of pressure from several quarters, not least, from Harald and myself, to release them. On 8 February 1979, the one-and-a-half-year-old Cathal was liberated and although he visited the food dump the next day, we never saw him after that. I suspect that, like Isla in May 1977, he failed to utilise the available food supplies and so must have starved over the next couple of months. In mid-August his skeleton was discovered on the beach at Sanday, Canna. Strangely enough, in a bizarre quirk of currents and circumstance, it was at the exact spot where Isla had been found a year earlier. Loki had also succumbed two months after her release, although perhaps not through natural causes, and these three ringing recoveries confirmed the hazards which the youngsters have to face early in their freedom.

The second long-term captive, Sula, was given her liberty on 6 March 1979, while Beccan observed the event from the sky above. As soon as Sula took to the air, Beccan swooped, forcing her to flip over and show her talons – quite an accomplishment on her maiden flight and after nearly three years' confinement. After several minutes in the air she landed on a grassy slope with Beccan at her side. The latter took off, circled and returned two or three times but Sula refused to budge, twice jumping at her aggressively. The one-year-old female Ulva suddenly appeared on the scene so Beccan pursued her instead; momentarily they touched talons in mid-air before both alighting beside Sula. When Sula did finally take wing, Beccan followed and coerced her into touching talons, perhaps in threat. This pair, though, were in their third year and perhaps courtship was in a frustrated Beccan's mind. It crossed my mind that I had perhaps sexed him correctly after all and that he was a really big male.

Sula landed beside a small pool where she proceeded to bathe. Beccan sat patiently nearby until forced to take wing by a persistent Ulva, whom she chose to ignore. Eventually Sula emerged from the water to dry off on a nearby knoll. I found her at exactly the same spot the next day but by the third day she was testing her flight capabilities in the strong onshore breeze. Several Hooded Crows mobbed her and once even succeeded in tugging momentarily at her outstretched wings.

It was at this time that the injured Shona was placed on tethers where she quickly mimicked the example of the still-captive male Ronan, even attempting to flutter

Male Ronan prior to his release from tethers, Rum, 1979

weakly on to the roof of her shelter. On 24 April 1979, once we were sure that she had settled on her tether, I released Ronan. Once or twice during the ensuing weeks he returned to feed alongside Shona but obviously rejoiced in his new-found freedom. He ranged extensively and by early June was seen regularly near the beach at Kilmory in the north of Rum. We found several gull carcasses there and Ronan also seems to have been responsible for killing two Eider as they innocently incubated eggs. Ronan had developed his hunting skills at an early stage, even before his release from tethers, for it was he who snatched, killed and consumed a Hooded Crow that had foolishly strayed within his reach.

Ronan's notoriety became firmly established when, on 23 June, he killed a red deer calf. It had been born two days earlier and had just been marked and weighed by Fiona Guinness as part of her study in Kilmory. She was observing from afar until the mother returned. After a heavy squall an eagle emerged from the hollow where the calf had been lying. Upon investigation Fiona found the calf dead, its ribs opened up with one foreleg torn out and eaten. The carcass remains weighed 5.2 kg, some 1.8 kg less than when it had been weighed only a few hours earlier. Ronan sat on a nearby hillock with a bulging crop. While in captivity he had weighed about 5 kg himself, so at this single meal he had consumed nearly 40 per cent of his own body weight. The deer calf had been healthy and of average weight, although its mother (herself tagged) was known from previous years to be particularly inattentive and a rather poor parent. Fiona had been waiting some distance away to make sure that the mother accepted the plastic collar and tags, for if she had not, they would have had to be removed. Now all Fiona could do was skin the calf's carcass, which bore no talon marks nor any signs that the victim had struggled; the wound where Ronan had grabbed its prey and fed must have been instantly fatal.

The Nimrod flies over Bliksvaer on the way from Bodø back to Scotland

On 20 June 1979 I flew to Norway to collect six eaglets. They proved to be three of each sex, though their ages varied from six to eight weeks. All fledged successfully and, with veterinary clearance, the first two males, Fillan and Ossian, were freed on 24 August. Within four days they were returning to feed at a deer carcass I had left out nearby. Sula later teamed up with them and all three were seen feeding at a deer gralloch on the open hill some distance away.

The author with a Sea Eagle chick, Rum

The remaining four juveniles were liberated during September, the male Conall on the 6th and the females Gisla and Eorsa on the 13th and 19th. They all established a routine of coming to the food laid close to where the last female, Tolsta, was still being kept; she was freed on 27 September. Both Ronan and Sula 'adopted' the juveniles, assuming the role that Beccan had played in previous years, her visits now being infrequent. On one memorable day I watched seven eagles in the air together. Several attempted to talon-grapple, but only the pair of older birds could successfully interlock, cartwheeling out of the sky with excited screams. Both Ronan and Eorsa were seen to pursue ravens from the food dump and pirate scraps from them.

On 18 September 1979 the invalid Shona died so there were no longer any tethered birds to act as decoys. As a result, the 1979 releases forsook the food dump earlier than had those of previous years, but at least they were still in the company of an older, more experienced pair. They were seen to locate grallochs and to feed at a deer carcass, and on a dead seal washed ashore. I shot and skinned three goats for them but these were ignored, suggesting that there

Juvenile Sea Eagle, Rum

Fully fledged Sea Eagle tearing up prey

Sula sitting on top of a cage, Rum, 1979

might already have been an abundance of carrion available. It was the height of the deer cull with plenty of grallochs lying around. The NCC stalkers Geordie Sturton and Louis Macrae would occasionally feel that they themselves were being stalked – by eagles waiting for them to shoot another deer. Once or twice the stalkers or their ghillies returned with ponies to retrieve a carcass, only to find an eagle sitting on top of it. The stalkers, though, took this in good part and continued to provide me with venison scraps as eagle food and excellent sight records from all over the island.

During the summer Sula acquired an oily patch on her breast, presumably from having killed a Fulmar chick head on. Normally Sea Eagles would attack adult Fulmars from behind as they took off from the sea, so were in no danger from any oil spat by the Fulmar. Her diet of carrion and her constant preening eventually allowed her to clean herself up, though, and she now appeared firmly attached to this corner of the island, whereas Ronan was back on the prowl in Kilmory Glen, the deer study area. He happily accepted some fish which I left on the beach for him but was mainly hunting gulls. On one occasion I watched him launch a surprise – but unsuccessful – attack on a flock of gulls roosting on the shore. He also fed at the carcass of a stag, cleverly evading my attempts to photograph him from a hide I had placed nearby. At times he would move along the coast, even being seen over the village of Kinloch, much to the delight of local residents and visitors.

During the summer months it was apparent that Sula and even Beccan were taking up territories on sea cliffs. Soon Ronan appeared to have paired up with one of them. These were encouraging signs indeed.

On 10 January 1980, the very pale-plumaged Gigha, who had fledged in 1977, reappeared in Rum, having apparently spent the summer on Mull, 35 km to the south. In mid-March she still accompanied Sula, who in early April entered her annual moult. Twice during that spring her footprints were found on the sandy shores of two freshwater lochs where she had presumably bathed. One or two reports of the 1979 releases were received during the summer. Perhaps the most extraordinary was that of the Sea Eagle seen by my

Harald Misund and the author collecting a Sea Eagle chick from a nest of twins

colleague Nigel Buxton near Tolsta on Lewis on 26 June. Being able to approach the bird closely, he could distinguish the two red rings on its left leg. Imagine my astonishment when I consulted my notes to find that the eagle Nigel had seen was the very one that I had named Tolsta.

On 20 June 1980, Martin Ball and I received eight eaglets from Harald Misund in Norway, the collection having been covered by the BBC. Hugh Miles stayed on to film adult Sea Eagles at the nest so another crew met us at Kinloss, coming to Rum to document their arrival. On 27 August I released two males, Cowal and Erin, so that when Hugh reached Rum he was able to film them from a hide at a food dump. As I have described, Hugh also shot the release of a female, Morna, on 8 September, and that of the male, Appin, the next day. Not to be left out, Sula performed for the camera too. The completed film, with some fine sequences of Sea Eagles catching fish in Norway and a commentary by David Attenborough, was screened a year later.

The female Hynba and the male Lorne were set free on 16 September and I retained a female, Croyla, and a male, Bran, until 2 October. Hynba was the more ready to use the food dump and on 10 October I found her perched confidently on the shelter to which a month earlier she had been tethered. Two weeks later I watched her approach the food dump, which that day contained a Great black-backed Gull which I had shot earlier. Although interested in the gull, she was reluctant to come near it. An hour passed before she eventually approached, cautiously. After several minutes she jumped tentatively forward but only brushed it with a talon, as though to ascertain that it really was dead. The next time she swept it up and, several metres away, finally plucked and ate it. On several other occasions I had been aware that some eagles displayed similar caution on approaching a dead bird, preferring if possible to accept meat or fish instead.

On 27 October Hynba had a bath in a shallow freshwater pool near the shore at Harris. She flew to a nearby fence post to dry off and preen, but first balanced precariously for a moment on the actual fence wire.

Early in November the pale Gigha again reappeared, having been absent just as she had been in previous summers. On 24 November I observed eight eagles at Harris and, as was to be expected, the interactions during the ensuing weeks were many and various. On one occasion I was watching several of the juveniles near the shore. One was flying lazily along the raised beach where it was stooped at by a bold Hooded Crow. Absent-mindedly the eagle flipped over to threaten with bared talons but sacrificed so much height that, being only 2 metres up, she crashed to the ground in an untidy and ignoble heap. Having regained her composure, the embarrassed eagle vented her annoyance on one of her peers perched innocently nearby.

During 1981 most of our sightings were still from Rum, with a few from the other Small Isles, the nearby mainland, Mull to the south or even the Outer Hebrides. Of 18 reports of Sea Eagles outwith Rum that year, one had ranged to Fair Isle in Shetland and another to the Antrim coast of Northern Ireland.

In June 1981 we learnt that the RAF had to bring forward our flight which, together with a poor and disrupted breeding season in northern Norway, resulted in Harald's only being able to collect five eaglets. Three were almost fledged while the other two were barely half-grown, but all were reared successfully and the two youngest – a female, Grania, and a male, Mabon – were the first to be liberated, on 18 September 1981. Another male, Nechtan, followed eleven days later and a female, Forsa, with the male named Brechin were released in mid-October. Sula had spent all year in the vicinity and, as if on cue, appeared as soon as Mabon had taken wing. She then remained with the juveniles near the food dump, when three other eagles released in previous years also turned up.

The first, possibly the female Gigha, was nearly adult, while another was established by its red ring as a male set free in 1978. The third I was able to identify positively as the female Hynba from 1980. It was especially gratifying to know that the nine eagles chose to remain in Rum that winter. We received reports of others elsewhere but sadly, in late April 1981, we had learnt that the male named Erin, released on 27 August 1980 and soon after filmed by the BBC, had been found dead in Caithness. Analysis of his carcass revealed that he had been poisoned, having fed on a hare carcass laced with the lethal poison Phosdrin. The bird had otherwise been in good condition and was of a healthy weight. In November 1981 we were notified of our fifth recovery, from the northwest of Skye, and perhaps the greatest loss of them all. The bird was a female, Vaila, which had been released in 1977; we had received no news of her since and she was by then almost old enough to breed. She had been dead several months so that little remained of her carcass to permit analysis, but a dead gull lay nearby and both birds lay a short distance from the carcass of a sheep which had been pulled to a conspicuous position on top of a knoll. We can only suspect that Vaila had been poisoned. At least two other Sea Eagles were seen in the area so, as a wise precaution, our contact buried the sheep.

Sula perched on a post near the tether site, Rum

A sixth recovery came in May 1982. The skeleton of the 1980 female Hynba was washed out from a ditch in Canna during a period of prolonged and heavy rain. She had been released on 16 September 1980 and had remained dependent upon the food dump throughout that and the following winter. In December 1981, during severe and heavy snow conditions, she disappeared. Hunting may have become difficult for her, especially as I was unable to replenish the food dump because my access road was blocked, and she seems to have moved to Canna. It is possible that, in a weakened state, she leapt into the ditch to feed on a dead sheep and was unable to get out, a technique that, as we have seen, was used to advantage in the Highlands to trap and kill eagles last century. Perhaps, though, she had simply starved to death in the cold weather.

On 22 June 1982 ten eaglets were flown to Kinloss by the Norwegian Air Force. All fledged successfully and, while in captivity, served to attract back to the vicinity the 1976 female Sula, in the company of a 1979 male and the 1980 female Croyla. The female Petra and the male Merlin were released in mid-September and utilised the food dump for several months. The others – six more females and only two males –were set free over the next three weeks. These eight were fitted with patagial wing tags, orange on the right and red on the left,

Young Sea Eagle with a patagial wing tag, Rum, 1982

each of which bore a conspicuous digit ranging from 0–9. Although rather obtrusive, these tags greatly facilitated individual recognition.

A male, Gregor, bearing the number 6, immediately forsook the vicinity and turned up six days later, some four miles to the northeast, on the shore at Kinloch and near my house. He was later seen at the north of the island, ending up on 14 October in the Isle of Muck, some six or seven miles to the south. Almost immediately he found himself swept far to the north during a period of easterly gales. On 19 October Nigel Buxton again contacted me to report Gregor's presence north of Stornoway in Lewis. Tagged Sea Eagles were seen both in Eigg and Canna in November but at least three still remained in Rum, making the most of deer grallochs and carcasses.

1983 proved an exciting year with our first eggs. We received a further ten eaglets from Norway, Sula visited the newly-occupied cages regularly and, as we have seen, brought the fresh young imports a dead Fulmar. The eagles were released during August, all of them sporting the usual leg rings but also yellow numbered wing tags.

Sula sometimes permitted a close approach, Rum, 1982

For a time, Beccan associated with the four birds still awaiting release from tethers. Some birds would later beg food from him. Most utilised the food dumps for several months. On 11 and 12 September the male Oran was seen at Kilmory in the north of Rum and three days later he was on the shore at Kinloch, five miles away. The male Fergus was found dead at Kilmory on 21 October 1983; he was our seventh recovery out of 62 eagles released so far. I was now living on the mainland and my notes became less detailed and more sporadic but it was obvious that, for some reason, we were getting fewer reports of this year's birds in and around Rum, but still quite a few of last year's. One of

1982's males, Gregor, remained near Stornoway until March 1983, once or twice venturing further afield but still within Lewis. In March and April 1983 two red-tagged (1982) individuals were reported in Orkney and in Shetland and another in Caithness during the first week of October. Older individual Sea Eagles, both immatures and near-adults, were sighted in Skye, in Sutherland and elsewhere.

Another ten were liberated in Rum in early August 1984 when, on one day, I freed six at once. This helped the eagles find each other quickly and team up to return almost immediately to the food dump. The remainder flew on 14 August, when the female Daltra was filmed for ITN News, and 16 August. The fewest sightings were made of the male Dulsie, the first to be released, and of Lethen, the last. On some days all ten were seen but the adult birds from previous years were much less in evidence. The male Clunas was from the outset most inclined to wander, being seen on 14 September when he harried Ravens to force them to drop scraps which he then retrieved. But the weather proved dry and sunny and several of the eagles were seen 'sunning' themselves, perched very erect, wings held open in heraldic pose and beak open, panting to cool off. Two were seen bathing in a shallow pool on 24 September. Female Clava remained dependent on the food being provided, sometimes in the company of Ronan. Although the food dumps were maintained, most juveniles were independent by the end of October and began to be spotted further afield. Four of them were gathered in Shetland for a time. Happily, no deaths were reported in 1984 nor in the following year.

Early in 1985 one red-tagged eagle (from 1982) was spotted in Northern Ireland. Once again I ringed, tagged and released ten fresh imports during August, six of them in a single day. Three of these birds were soon to be joined by a four-year-old sub-adult. It proved to be a very wet summer so sightings were a bit sparse. Phase I, as it came to be known, had officially ended and no more birds were released in Rum.

Sea Eagle being mobbed by Ravens, Rum

On 15 February 1987 a female Rona, tagged and released in 1984, was found freshly dead in Torridon on the mainland – our eighth recovery. In September 1988 the BBC repeated the *Wildlife on One* programme on national television. I produced a fully illustrated summary of the project so far on behalf of the team, to be published in November 1988 as *Research and Survey Report No. 12* (Love 1988). I also began assembling a computer database of all sightings received since the project began in 1975, which soon exceeded 3,000 in total. It seemed an opportune time to assess the progress of the project:

- Eight of the 82 Sea Eagles released in Rum had so far been reported dead, a mortality of 10 per cent. Four of these had died in their first year and another, Cathal, who had been retained on tethers for two and a half years, died within six months of being freed.
- A four-year-old female (Vaila) and a first-year male (Erin) were victims of illegal poisoning. The only other reported cause of death was that of a first-year female (Loki) who had apparently come to grief on overhead power cables, although it is now suspected that she too was a victim of illegal persecution.
- The five remaining deaths occurred in unknown circumstances. The birds may have been unable to adapt to the wild; two of them had failed to utilise the feeding stations upon release and a third was seen there only once. On their release, all but one of the eight dead birds had been 1–14 per cent below average weight. The timing of their release may also have been implicated in mortality. Eight sea eagles were freed during the spring months and two of these (25 per cent) were later reported dead, contrasting with only six deaths amongst the 74 autumn releases (8 per cent). Perhaps the lengthening summer days encouraged the spring releases to quit the vicinity of the feeding station too soon, at a time of year when alternative food supplies, in the form of carrion, were scarce. It was accepted that other deaths must have occurred undetected but, nonetheless, estimates of overall survival suggest that these have been few.

At first we relied upon casual records but, by 1981, surveys outwith Rum could be increased because of funding obtained for RSPB field workers. Up to that time I had amassed 969 sightings, either locally, on neighbouring islands or on the adjacent mainland. Not surprisingly, about 75 per cent were made in Rum itself, mostly in autumn and winter. By spring 1982 Sea Eagles were tending to be recorded 30–60 miles away, largely as a result of better surveys but also because of particularly stormy weather the previous autumn. The most distant sightings occurred in the summer but, up to 1981, some 97 per cent of Sea Eagle sightings were still within 60 miles of Rum.

This pattern changed little over the next seven years as birds attaining breeding age established permanent territories. Perhaps it is interesting to note that the eagles which were wing tagged in 1982 tended to disperse further afield than any released in the final three years of Phase I. Not only were the 1982 releases made one or two months later than normal but, as in the previous year, it was also a very stormy autumn. One individual (Gregor) was recorded in Rum on 13 October but turned up over a hundred miles to the north only six days later, after a period of prolonged southeasterly gales.

The distribution and timing of these early reports provided several opportunities to estimate survival. Although colour rings were rarely identified, wing tags, plumage features and moult patterns facilitated the recognition of individuals in widely- scattered localities. By January 1982, for example, at least 20 separate individuals out of 42 then released (none yet tagged) could still be accounted for. Similar estimates were made early in 1984 (62 released) and early in 1985 (by now 72 released, 38 of them tagged). These estimates suggested a minimum survival of about 60 per cent. Since by 1988 only eight had been found dead, survival – theoretically at least – could be as high as 89 per cent.

Sightings of the wing-tagged birds (the colour revealing the age cohort even if the bird was not individually recognisable) provided me with best estimates of survival. Eight or nine of the ten released in 1982 were still alive over two years later. Amongst a further ten released in 1983, seven or eight survived at least until the following spring, as did at least eight of the ten 1984 releases. It is difficult to give similar information for the final ten, tagged in 1985, as no captive birds were then present in Rum; previously these had served to attract young eagles back, allowing sightings of their tags. However, of the 38 tagged eagles for which adequate data was available to me in 1988, as many as 80 per cent lived through their first year at least, the period during which they are presumably most at risk. As more and more eagles were liberated they tended to congregate at favoured feeding locations or at roosts, whilst older birds became available to assume the role of foster parents.

A consideration, finally, of diet. The high survival rate of the Sea Eagles is firm proof that they have found their habitat perfectly suitable, with adequate, perhaps even abundant, prey. Casual observations accumulated over the years have confirmed the importance of carrion in their diet, especially in winter. Alan Leitch analysed 127 pellets from winter roosts of young Sea Eagles in Mull. No fewer than 166 prey items were recognised, 40 per cent of which comprised carrion in the form of deer, sheep or goats. There was a surprising preponderance of small and medium mammal prey (48 per cent), mostly rabbits and hares. The remaining 12 per cent were bird prey including two grouse, two gulls, three ducks and eight seabirds (Watson, Leitch and Broad 1992). These Mull roosts lay some two miles inland and much of the area over which the Sea Eagles hunted was, of course, deficient in seabird prey. Furthermore of course, mammal fur can be over-represented in pellets at the expense of bird and especially fish remains (Chapter 4).

A close-up of an immature female in her third year, Rum

On the other hand, when prey remains are examined at nests, birds tend to be overestimated. A preliminary analysis was made (Watson, Leitch and Broad 1992) of 114 such remains from the west of Scotland, together with sight records of eagles with food, mostly in the summer months. This showed 51 per cent to comprise bird prey, mostly Fulmar, gulls and ducks, but including auks, grouse, Shag and Hooded Crow. The mammal prey made up 27 per cent of the prey items, again mostly rabbits and hares, and carrion a further 16 per cent. The remaining 6 per cent were fish. There are, in addition, a few reports of fish being caught. One Sea Eagle became something of a local celebrity by subsisting for several months on fish baits thrown out for it by lobster fishermen. This is now a common occurrence. It is likely that fish is still grossly underestimated in these studies of our Sea Eagles' diet.

One of the young Sea Eagles succeeded in snatching a Herring Gull that was mobbing it, while another was observed appropriating a fish from a Heron. Roy Dennis watched one on Loch Torridon attempting to steal fish from an otter with cubs. These sea-going mammals are known to provide an important source of freshly- caught or half-eaten fish in Norway and now they do the same in Scotland too. Rum, like all the Hebrides and the west coast mainland, has a good population of otters.

By 1985, having released a total of 82 young Sea Eagles (43 females and 39 males) the project was now into a whole new ball game: the establishment of breeding pairs in the wild. The initial steps had been made in 1983, a couple of years before, so here we have to retrace our steps a little.

Chapter 10
The native has returned

'I hear the eagle bird
With his great feathers spread,
Pulling the blanket back from the east.
How swiftly he flies,
Bearing the sun to the morning.
Iroquois poem

The Sea Eagles released in Rum in 1975 were expected to mature in five years or so. While activity did intensify as expected, initial developments were slow, the delay undoubtedly due to the fact that in these early years there were still relatively few birds roaming around a large vacant area.

In Rum, Sula was well established by 1980 and I frequently saw her in the company of various immature birds. I even constructed a crude nest for her on a cliff overlooking the tethered site but, quite justifiably I suppose, she ignored it. It did serve as an occasional roost for the other Sea Eagles while Sula took an interest in a recently-vacated Golden Eagle nest at Papadil in the southeast of Rum. This pair of goldies had always had a poor breeding success and it may even have been the increasingly frequent presence of Sula and her young consorts that induced them finally to move away.

As a species the Golden Eagle is the more aggressive and in the early years of the project I frequently witnessed their attacks on Sea Eagles. Once a dive was so determined that I feared for 'my' Sea Eagle's life. Any closer and the Golden Eagle would have ripped open her skull with its claw. Interestingly, though, while I had witnessed at least one attack in each of the first 18 months since the liberation of the first Sea Eagles, I was now witnessing only one a year. Since the aggression was having little effect anyway, it seems that the Golden Eagle had to become more accepting of its new neighbour. Two of Rum's three pairs of Golden Eagles soon moved to new eyrie sites, one well inland, where, being the 'mountain eagle', it more properly belonged.

I had been witnessing more and more aerial display and calling and, on more than one occasion, stick carrying. On 20 August 1980, on the remote western cliffs of Rum overlooking Canna, I disturbed an adult from a ledge where I had earlier found some sticks. Although its

In good weather, courting Sea Eagles may spend some time in the air, soaring around together almost wing tip to wing tip

An adult Sea Eagle in full display

beak was full yellow and its plumage the characteristic pale greyish-brown, some of its tail feathers retained a few dark tips. It was a male and in heavy moult. Suddenly, to my left, a second bird appeared, very much larger and obviously an adult female. It seemed to be of a similar age but in a more advanced state of moult, its tail almost fully regrown and quite white. Both soared around a bit before disappearing. On my next visit I was delighted to make out the short leather jess on the female, confirming her to be Beccan, while the colour rings on the male revealed him to be Ronan, both 1976 birds that had been held in captivity for

Bloodstone Hill near the last nest of Sea Eagles in Rum in 1909

some time before being released or, in Beccan's case, escaping. They were using several ledges as plucking posts and on them I could make out the remains of young fulmars, auks and gulls.

During the winter of 1980–1981, aerial courtship intensified and in the following spring, cruising past in the Rum boat, I could make out a structure of sticks above me that had been invisible from the cliff top. I eventually located a spot near the shore from where I could view the nest, but three prospecting fulmars were sitting on its edge so the eagles were no longer interested. It was in this area, perhaps on this very ledge, that the last Sea Eagles in Rum had nested in 1909.

My expectations for 1982 were high but for some reason Beccan and Ronan were rarely to be seen and their nest showed no signs of recent occupation. I suspect they had moved across to Canna and we now know that a pair can include both islands – four miles (6.5 km) apart – in their territory, with alternative nests on each. Sula in the meantime continued to occupy the equally remote southeast corner of Rum, although I had little inkling that she had found a mate.

With Sea Eagle activity intensifying on Skye, and on Mull in particular, it was obvious that surveying and monitoring needed to be stepped up. By the late 1970s, Richard Coomber was running a guest house in Tobermory on Mull, also offering wildlife tours around the island. He began notifying me of more and more sightings so it was obvious that this was going to be a key area. The RSPB were now well placed to take on the bulk of this role, although I continued to wander the lonely hills of Rum and the lovely island of Canna.

Roger Broad, RSPB's Regional Officer for Argyll

As the RSPB's new Regional Officer for Argyll and the islands, Roger began to mobilise better cover in Mull. Various field workers were deployed in Mull – Dave Sexton, Mike Madders, Keith Morton and others – with Colin Crooke and then Kate Nellist and Ken Crane, Justin Grant and others on Skye. In 1981 the team had identified at least 15 of our Sea Eagles in Mull. These birds hunted on the coast but were also seen to be especially fond of two particular inland areas. A small conifer plantation began to be used as a roost by several immatures, a rather pale-backed two-year-old male and a three-year-old female, until they were scared off by timber operations. After discreet approaches to the foresters, further felling was suspended. Two immatures remained together as a pair and, on occasions, were seen at the other inland site with a juvenile and four mature Sea Eagles. Three of the latter were in full adult plumage and one had a black leg ring indicating that it had been released in 1975, so it was either Karla or Freya.

On 29 May 1981 two of the adults were seen copulating on a rock ledge in Mull and the following spring indulged in further aerial display and talon-grappling. At times a four-

Sea eagles talon-grappling, Rum

year-old female would interfere and was once seen attempting to talon-grapple with the young pale-backed male.

During April and May 1982 adults in Mull had even been seen carrying sticks to a cliff a few miles inland. This ledge was nearly 5 metres long and 1 metre wide and soon contained a rather crude nest under an overhang, which became littered with droppings, down and food remains. Finally, at the end of May, during a spell of very wet weather, the ledge became waterlogged and was abandoned. It was rather an accessible site so we were relieved that they had abandoned it. In the meantime a third adult had been carrying sticks to a ledge on the coast of Mull, while two more adults were seen further inland. A single adult had also taken up residence for a time in Glen Affric on the mainland, where it occasionally displayed at the local Golden Eagles and even constructed a crude eyrie of sticks on a ledge, but this came to nothing.

The coastline of southeast Mull

Mull really became the focus of attention for the next few years. By 1982 it seemed that three pairs might have become established there. By April 1983 one pair had come to occupy a cliff ledge inland, only a mile or two from the sea. Then, though, the male began to show a lively interest in a second, dark-headed adult female. The male had even copulated with her on a ledge about 200 metres away and she was soon sharing incubation. On 9 April 1983 the original pale-headed female was seen trying to block her rival's attempt to sit on the nest. The determined challenger edged past her, however, and with careless enthusiasm jumped into the nest cup to assume incubating. It could have been at this moment that the clutch was damaged and perhaps by this time the rival had already succeeded in laying her own two eggs in the nest. This uneasy triangle of the male with his two consorts persisted until he ousted his original pale-headed mate completely. The nest was examined on 14 April and found to contain two intact eggs, together with the fragments of at least one other. By mid-May, however, after three days of continuous low cloud and rain, the eyrie was found to be deserted.

The cliff in southeast Mull where the first eggs were laid, April 1983

The remains of two clutches in the first nest located on the cliff ledge in Mull, May 1983 (Roger Broad)

In the meantime, on 29 April 1983 a second, very anxious pair were found to have a nest, albeit a poor structure in a leaning tree in a copse near the cliff trio. The pale female still displayed dark tips to her otherwise white tail so was not quite yet fully adult and probably four years old. Enter Blondie, the bird which Dave Sexton would later dub 'one of the most famous Sea Eagles in the world'. Her mate, however, showed much less white on his tail and, with a speckled back, was apparently a year younger. On 2 May one bird was brooding and the nest was found to contain a single egg. Blondie was seen to be incubating but by 15 May she had abandoned the egg, which was later found to have been broken. Unlike the other two clutches on the cliff ledge, the shell of this egg was rather thin-shelled.

Thus, in mid-May, both pairs failed in their breeding attempts and it may have been mere coincidence that this was during a spell of bad weather, with heavy rain and storms. In neither case could the circumstances be considered normal. Although it resulted in two clutches, the

The eyrie constructed in a tree by the second, young Mull pair, April 1983

The single egg deserted by the young pair in May 1983, Mull (Roger Broad)

first was an awkward triangle of the male with two females, while in the second case both birds were not quite fully mature, the male seemingly a year younger than the female; they had chosen a precarious nest site and had constructed a rather poor nest in which their first egg proved slightly thin-shelled, all doubtless due to immaturity and inexperience. It is unfortunate, too, that all these first breeding attempts were made in a season that proved a poor one for golden eagles elsewhere, not just in Mull but throughout the Highlands. A major landmark had nonetheless occurred, with the first proven breeding attempts in Britain for seven decades.

At the third locality in Mull, a pair of adults had been seen for two years, but late in March 1983 the local people noticed the birds had become more secretive. An adult flying in with a stick led to the discovery of a large nest in an oak tree, perhaps even built the previous year, but nothing was to come of it.

So 1984 arrived with a distinct air of anticipation within the project team. Mike Madders and Dave Sexton, working for the RSPB, found the trio still in possession of their cliff eyrie but no evidence of a breeding attempt. The young pair, meanwhile, had built a new and better structure in a tree some distance from the previous year's one. They diligently incubated a single egg for seven weeks but it failed to hatch. This time its shell was found to be of normal

The new eyrie constructed in an oak tree by the young pair in spring 1984

A mirror is hoisted on a pole to reveal a single egg from the young pair in April 1984; sadly it did not hatch

The location of a new, remote cliff nest found in Lewis in the Outer Hebrides in 1984

The first chick fledged in the UK for 70 years in southeast Mull, May 1985 (Roger Broad)

thickness and its contents were analysed. Traces of pesticides were detected but nothing to cause undue concern. Blondie's mate was still not quite fully adult and it seems that, once again, age and inexperience could have accounted for the egg's infertility.

Towards the close of 1983 Andy Miller Mundy had contacted me about a nest he had found in the west of Lewis, in the Outer Hebrides. In late March 1984 we visited the site and, from a safe distance, discovered the pair to be incubating. Within a fortnight, however, during a period of cold weather, the nest was abandoned and in early May we found it empty, the egg or eggs presumably having been taken by crows. A few other potential pairs elsewhere were showing encouraging signs, including a young pair found by Ken Crane and Kate Nellist that had constructed a crude nest, which Roy Dennis later rebuilt and improved, in a conifer plantation in Skye. In the end, though, only two clutches were laid during the 1984 season – one from a 'new' pair in Lewis – and, disappointingly, neither had hatched.

Dave Sexton and Mike Madders Mull, 1985 (Roger Broad)

Maybe 1985 would prove more successful. The trio had finally split up. The male moved with his second female to a new site, where they built a substantial tree nest. This was, however, too late in the season for any eggs to be laid. The young pair had now both matured and by late March were confidently sitting on a clutch of two eggs. We suspected that the male had been released in 1980 and the female the year before. Both eggs hatched in early May but one chick died almost immediately. The other thrived, however, and fledged twelve weeks later to become the first Sea Eagle bred in Britain in seventy years, a hugely significant milestone for the project.

Blondie had come up trumps. Dave Sexton, one of the RSPB wardens on Mull, remembered his excitement:

> It is one of those moments in wildlife conservation history that can never be repeated… Being there to watch the first white-tailed eagle chick hatch and fledge in the UK for 70 years was thrilling, dramatic, stressful and exhausting – all at once! That momentous occasion was in 1985, a full ten years after the main Scottish west coast reintroduction project commenced on the remote Hebridean Isle of Rum… I've been in awe of this bird since the main UK project began. I saw my first White-tailed eagle on a holiday to Mull in 1980 and was desperate to work with them. I achieved my dream four years later with a contract from the RSPB to protect those first few pioneering pairs attempting to nest back in Scotland – their native land. And then that moment, a year later, when I and my colleagues Mike Madders and Keith Morton confirmed the hatch. Their CB radio

> crackled into life. The signal was poor but the message was clear: 'I think we're both daddies!' Then, 12 weeks later, the fledging of that first chick! I remember it as if it were yesterday.

Dave continued:

> Blondie hatched her two eggs one mild spring morning and offered those precious, historic chicks their first meal. I can still see that big yellow beak turning left and right as she carefully moved her head to get the meaty morsels in the perfect position for the tiny beaks to accept them. Her massive talons safely clenched as she stalked around the nest rim and then settled back down to brood the downy bundles of joy. Ah, what a girl. She looked like she'd been doing it all her life but this was all new to her. Her inexperience eventually showed when we lost one of those first chicks at about three weeks old. Maybe not enough food came in, maybe some sibling rivalry?
>
> All I know is we watched that remaining chick with an almost obsessive intensity, noting every move, every feed, every squirt. Nervous days and nights to be sure. And then following that longed-for maiden flight, what did Junior do? He gave us all heart failure as he ditched in the middle of the loch, floundered about and then vanished without trace. Was he dead or alive? No one knew until the following morning when we found him sitting hunched and bedraggled at the loch edge with Mum Blondie and the male either side of him for support. From then on he went from strength to strength and did what young white-tailed eagles are meant to do.

This juvenile would be seen in Mull late into 1986 but even now Dave cannot stop thinking about him:

Sula at her nest in southeast Rum, revealing the single egg, April 1985

> I so wish we knew where he is today. We didn't wing tag him in those early days so after he dispersed from the breeding area, we only got occasional sightings of a 'young untagged sea eagle'. I like to think he eventually settled down years later and bred successfully himself. Knowing how long they can live, he could even still be out there today. I think he is.

In Lewis in 1985 our third pair had laid again and – with thanks to Andy Miller Mundy – NCC was able to install Mike Cook in a lonely cottage nearby so he could monitor progress in all weathers! From a nearby clifftop he could watch the nest with a telescope and gallantly gathered a lot of useful observations on behaviour at the nest. But the very day the eggs hatched, an unexpectedly severe snowstorm occurred and the two young succumbed. This eyrie faced north, as did a new one elsewhere which also contained eggs, and both received the full brunt of the storm. The new pair abandoned their eggs.

Meanwhile, at a remote corner of Rum, I was proudly following Sula, at last incubating a clutch of her own on a steep cliff where her kind once bred in the distant past, perhaps on the very same ledge. On one occasion I watched her stand up to reveal one egg in the nest cup; a few days later she was sitting up just as she might if she had a small chick tucked into her breast feathers. I was convinced the egg had hatched but sadly it must have died soon after, for the nest was abandoned by my next visit.

Our elation at the success in Mull was tempered only slightly by the knowledge that, were it not for the notoriously bad summer of 1985, two other broods might also have fledged from no fewer than four clutches laid and six active sites that season. No more young were to be released in Rum and the next stage – breeding in the wild – was exhibiting a rather slow beginning, but understandably so. Only half of the 82 Sea Eagles released since 1975 had so far matured, the remainder having yet to achieve sexual maturity. We know of course that at least 10 per cent had not survived so we still had a relatively small pool of potential breeders, which were dispersed throughout the western seaboard from Lewis to Mull. While weather was obviously having an impact on the breeding success, inexperience was probably a more important factor. The lack of any established breeding population to set an example, or even with which to pair up, was offset by the empty habitat which seemed to encourage young birds to attempt breeding earlier than they might otherwise have done. We would not expect a young bird, let alone a young pair, to be able to breed successfully from the start. (In fact, a few young pairs would in later years prove us wrong.)

By late February 1986 Blondie and her mate were regularly roosting side by side on their eyrie. As in previous years they maintained a relatively low profile and were rarely seen in courtship display. Mating was observed on 17 March, and repeated shortly afterwards on a nearby hillside. One adult was seen to be incubating on 21 March, a week earlier than in 1985. Their first egg hatched on 28 April, 38 days after laying, and the other by 1 May. Both chicks, probably females, fledged successfully 12 weeks later, on 19 and 21 July. Some food was left near the nest which the adults occasionally utilised but, unlike the previous year, they were having no difficulty finding enough for themselves and for their twins. Both fledglings were seen regularly until 18 August when one disappeared; it was found dead in Skye nearly a year later, on 6 June 1987. The other remained near its nesting territory in Mull, where the single juvenile reared the year before was also being seen.

Although the trio had sometimes been seen together over the winter of 1985–1986 near their original cliff ledge in Mull, by late February the male and his dark-headed female began building up an alternative site where they were conspicuous in courtship and copulation attempts. By 18 March one of them was incubating and on 24 April, just as hatching was expected, it was obvious that something was wrong. The nest was abandoned the next day. It was found to contain an almost intact egg membrane with shell fragments adhering to a well-developed embryo which was collected for analysis (M. Marquiss, pers. comm.)

Meanwhile, by 24 April 1986, the Lewis pair were incubating two eggs in a new nest but abandoned the attempt by 14 May, when a single egg was recovered for analysis; it contained a half-grown embryo. Both eggs, especially the latter egg/embryo, revealed surprisingly high residues of PCBs (Table 5). The third Mull pair similarly abandoned their nest mid-May leaving a single empty egg, apparently holed by a crow. In all, seven pairs were located in 1986, with four or possibly five laying eggs, but in the end only one – the previous year's historic pair – succeeded this time in rearing twins. Blondie scored again!

Table 5: Pesticide levels in Sea Eagle eggs from Scotland (1984–1986)

	In Lipid			Wet Weight		
	1984	1986	1986	1984	1986	1986
Sample units	Egg ppm	Egg ppm	Embryo	Egg ppm	Egg ppm	Embryo
			mg/g			mg/g
Analysed by	ITE	ITE	J. Bogan	ITE	ITE	J. Bogan
BHC	1.27			0.06		
Heptachlo repoxide	1.59	2.85		0.07	0.29	
HEOD (from aldrin and dieldrin)	6.37	86.27	0.32	0.28	8.07	0.01
DDE (from DDE)	29.30	313.01	84.3	1.28	29.27	3.2
PCBs	115.61	344.21	700.3	5.08	32.19	26.82
Mercury		0.56				

No fewer than nine pairs were located holding territories in the 1987 season. Six of them laid eggs but four did not hatch. The Outer Hebrides pair deserted in mid-May, Skye now had four established pairs, the Small Isles two, with another pair of hopefuls in Lewis. The bigamous male in Mull returned to his original female and built up the tree eyrie that had been found in 1983 but never used. Incubation commenced as early as 9 March and a single chick, thought to be a female, was fledged in mid-July. Although Blondie, from the previous year's successful nest, had not been much in evidence over the winter, both she and her mate returned to the eyrie in the spring. They began incubating on 27 March, nearly a week later than the year before, and successfully fledged two young – thought to be a male and a female – at the end of July. Both nests had been rigorously protected by the RSPB and all three chicks were ringed and tagged before fledging. They were both fed by the parents up to ten weeks after fledging. As already mentioned, one of Blondie's twins from the previous year was found dead on Skye on 6 June, while the female Rona, released on Rum in 1984, was picked up freshly dead in Torridon on 15 February. Later, in August 1989, another of 1987's wild-bred young would be found dead at Acharacle, Lochaber.

It is worth pausing here to summarise our observations on breeding so far. These I have drawn up from my own notes and in particular from my initial analysis of the detailed logs kept by the RSPB wardens in Mull and Mike Cook in Lewis, so they refer only to the first breeding attempts in the mid–late 1980s. Thus far, of the twenty nests located by the end of 1987, five had been built in trees (four in oaks and one in a spruce); the remainder were on cliff ledges. Six pairs were known to have alternative sites: Blondie and her mate had both their nests in trees, the trio had a cliff site and two further tree sites, and the young pair in Skye, yet to lay eggs and which had built in the spruce, had spurned Roy Dennis's efforts and chosen a cliff site instead. The remaining pairs had all their nests on cliff ledges.

Such nest site selection differs markedly from Norway where Harald Misund and others find many eyries in trees; cliff sites are usually only chosen further up the fjords or on islands off the outer coast. Some of these remote pairs, presumably due to lack of disturbance, may even nest on relatively flat ground. If there are trees available, though – and they need not be tall ones – then they are preferred. Even on cliff ledges the nest is often braced against the base of a small tree. Scotland has a sad history of deforestation over the centuries and it is doubtless the lack of suitable trees that caused the reintroduced birds to opt for so many cliff sites. As time goes on, our eagles are spreading more and more into areas where trees occur more commonly, such as Argyll or Wester Ross.

At this juncture it might be useful to highlight some initial traits shown by the first of the reintroduced pairs to nest in the wild.

Blondie first laid eggs when she was four and her mate only three, perhaps stimulated into breeding condition by their proximity to the promiscuous trio. A male in the Small Isles may have been three when he first bred although his mate was probably nine. On the other hand, it was the male of the Lewis pair that was the older, not breeding until he was eight while his mate was only five. The trio were thought to have been six or seven when they bred while the fourth egg-laying pair were both nine. Such a wide age span of first breeding may be linked to the longer period spent in captivity, two of the oldest having been retained on tethers for three and a half years before they were finally released and another having escaped from its tether after a year. On the plus side, these birds were better hefted to the vicinity of the release so their first nests were more easily located.

The trio laid their first eggs between 4 and 6 April but their next (1986) attempt had commenced by 18 March. With the male mated to the other female the following year, the eggs appeared on 9 March. Blondie is thought to have laid her first egg around 4 April 1984, on the 28 or 29 March in 1985 and about a week earlier (21 March) the following year, but she would slip back six days in 1987 and so on. Age and experience both obviously influence egg laying, which normally becomes progressively earlier with each breeding attempt. With clutch size, single eggs are more likely to be laid by first-time breeders, with two being the normal clutch. Only later have we recorded a few three egg clutches in the reintroduced population.

I analysed all the clutches I came across, both in the literature and in museum collections, to ascertain just what the situation in Scotland had once been, and how it compared with Norway (Table 6).

Table 6: Clutch size of the White-tailed Sea Eagle in Britain and Norway

		No. of eggs per clutch				Mean + Standard Deviation
	Source	1	2	3	4	(Sample size)
Britain	Literature	3	5	4	1	
	Collections	25	29	4		
	Both	28	34	8	1	
Norway	(Willgohs 1961)	5	40	12	1	2.16 ± 0.59 (58)

Adult female Sea Eagle incubating, Burg Guttenburg, Germany

While three egg clutches were known here – maybe slightly fewer than in Norway – two was the normal clutch in both countries. The preponderance of single eggs in Scotland may have been a bias introduced by egg collectors who sometimes split up clutches to pass on to fellow enthusiasts. Equally, an egg may have broken on initial collection or later, back home. The clutch might also have been reduced to a single egg before the egger raided the nest. For all these reasons the average British clutch is slightly lower than that commonly found in Norway or elsewhere. Fresh eggs weigh 138–145 g and several authors mention how the second egg might be smaller than the first. In his captive birds, for instance, Fentzloff recorded a mean of 124 g from five second eggs, compared with 135 g from four first eggs. A replacement egg might be laid if the first was destroyed or removed experimentally; it could be smaller still and Fentzloff, who strove to incubate Sea Eagle eggs artificially, found three or four weeks might elapse before this could happen.

The incubation regime was fairly typical of the species as a whole, the longest recorded stints being undertaken by the female of the Lewis pair in 1985, two lasting four and a half hours and one a remarkable seven hours and ten minutes. Only once, on the other hand did the male sit for as long as four and a half hours. Willgohs estimated the male's contribution might amount to some 27 per cent, though this can vary between pairs. One of Fentzloff's captive males was unusually attentive and sometimes excluded the female for prolonged periods.

Fresh nest material might be brought in several times during incubation, mostly by the male. Fentzloff, however, noted how his captive pairs restricted the bringing of fresh pine

The first chick, only a day or two old, has emerged with the second egg yet to hatch, Norway (Johan Willgohs)

Adult Sea Eagle with a chick in the nest, Norway (Johan Willgohs)

By its fifth week the chick's plumage is well developed, Norway

sprays to the nest to only one or two days prior to laying and again around the time of hatching. This habit has generated some speculation as to its function – the provision of insulation or humidity during incubation, to cover faeces and carnage on the nest or to aid sanitation once the young have hatched. Recently it has been thought that sprays of pine in particular may help reduce infestation by flies and nest parasites.

Fourteen items of prey were recorded being brought in during incubation by the Lewis pair in 1985 compared with sixteen by Blondie and her mate in Mull. These usually initiated a handover by the sitting female to the incoming male with his offering. On only four occasions did the Lewis female remain on the nest to consume the prey, all during the first three weeks of incubation. Willgohs noted that his male brought in only four items of food, three of which the female then left the nest to devour. Two of the three occasions on which Mike Cook's female brought prey to the eyrie were immediately prior to the hatch. By this time, either the male was showing less interest in brooding (19 per cent) or the female was reluctant to leave. She also spent shorter periods off duty and on returning might coax her partner off by gently pecking at his neck. Both parents tended to become more agitated during hatching and would stand repeatedly looking into the nest cup, sometimes gently prodding the eggs with their bills. Thereafter, the duty bird would settle with an

A Sea Eagle chick about three weeks old, still covered in grey down, with the carcass of an eider duck beside it, Sweden (Björn Helander)

exaggerated rocking motion and then sit noticeably higher in the nest. In 1987 Blondie's mate remained in attendance nearby and was the first to be seen feeding the chick.

A newly-hatched chick weighs 90–100 g and is relatively helpless for the first week or so. For the first couple of weeks it is covered in white down and only able to crawl rather weakly around the nest cup. At first, the Outer Hebrides male brought in one or two prey items each day, which either he or the female would tear up to feed to the eaglets. On occasions both adults presented food simultaneously and they were even seen to masticate the food in their bills first. The eaglets were fed regularly and often, and without discrimination. One female of the former trio was reluctant to let the male feed the chicks and it might have been fully two weeks before he finally got his chance. Almost at once the female would take her turn at hunting, but the chicks' initial inability to thermoregulate demanded constant adult attendance at the nest, continuing well into the second week. Usually the parent only left the nest to see off intruders, especially crows.

By the tenth day the chick may begin pulling morsels from the parent's bill and by the third week the white down is replaced with grey down. By another week it begins standing up, rather shakily, on its tarsi, and is thus able to defecate over the rim of the nest. By its third week, the chick is being left for longer periods and from then on, daytime visits by the parents, even in poor weather, are only to deliver food and to feed the chick. Two weeks later it is stronger on its feet and feeds voraciously from the parent, as well as snatching food from the floor of the nest in an effort to begin feeding itself.

Sibling aggression was at first infrequent and brief but during the fourth week in 1985 it became evident that Blondie and her mate were having difficulty finding enough prey, perhaps owing to the persistent wet weather. Sibling rivalry became more prevalent and by 5 June – about a month after hatching – only one chick remained alive. Thereafter the RSPB wardens began to leave dead rabbits and hares near the nest, food that was readily accepted by the parents. As a precautionary measure food was again made available during this period in 1986. The adults again accepted it but owing to a poor lambing season there was no shortage of carrion so the chicks were well fed.

Although eaglets of this age are quite active, they lie quiet when someone is at the nest, here in the Outer Hebrides

A chick's first feathers begin to emerge from about the fourth week. By the fifth week the chicks become more active, wing flapping, preening and picking up small morsels on the nest platform. By this point the 1987 twins were old enough to be wing tagged, enabling reliable individual recognition. Eight days later it became apparent that the older (female) chick dominated the smaller male. She readily snatched food from her parents while her sibling cowered to one

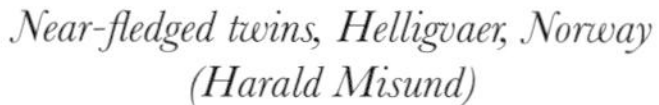

Near-fledged twins, Helligvaer, Norway (Harald Misund)

Some Sea Eagle pairs rear only a single offspring, Norway

side. The little male was, however, prepared to tear up and swallow the remnants once his sister was satiated. This is probably the normal situation so that siblicide – the larger depriving or even killing the smaller chick, what Seton Gordon termed 'Cain and Abel rivalry' – is relatively rare in Sea Eagles compared with Goldens. I had noticed a similar scenario amongst the captives, the smaller in reality being the more precocious and having to take food whenever it could, while it was the larger that often preferred to be fed by the parents.

By the seventh week the fledglings had begun to exercise their wings. Visits by the parents had become less frequent by the end of that week, when they were spending less than 10 per cent of their time at the eyrie. During the ninth week, with wing and tail feathers nearing full development, the fledglings indulged in vigorous wing exercises. In 1986 one eaglet even wandered along a stout branch, some 2 metres from the nest itself. Such tentative excursions by the single 1985 chick had not been noted for a further two weeks.

Johan Willgohs had judged fledging at ten weeks but most of our Scottish chicks seemed to fledge at 11–12 weeks. The 1986 twins fledged two days apart, while the following year the twins left this nest only one day apart. They all periodically returned to the nest for a further few days, and indeed the single chick from the other nest that year returned for a period of at least three weeks. Initial flights were short and tentative and the landings distinctly clumsy and awkward. Other than the twin that moved off, to be found dead in Skye the following year, one or two of the other Mull juveniles remained in the vicinity of the nest for several months.

About a third of nesting attempts in Norway and elsewhere might fail to produce any young, though up to 75 per cent failures have sometimes been recorded. The success of any population, or indeed of a single pair, may vary from year to year. Willgohs noted the average brood size in Norway to be 1.6 (from a sample of 160), while my own scrutiny of old Scottish records yielded a figure of 1.7 (15 broods). It was too early to deduce a figure for the newly-reintroduced population although my preliminary assessment suggested 1.6. About half of Norwegian broods consisted of a single chick but two is common, and Willgohs recorded four broods of three young and even one of four. I know of several instances from the past of triplets in Scotland and Ireland (Table 6) and recently we have recorded two or three instances in our reintroduced population.

Table 7: Brood sizes of some White-tailed Sea Eagle populations

No. of chicks/brood	1		2		3		4		Source
	No	%	No	%	No	%	No	%	
Britain and Ireland	6	40	7	47	2	13			Love 1983
Norway	39	42	49	53	4	4	1	1	Willgohs 1961
Poland	13	28	33	70	1	2			Banzhaf 1937
East Germany	9	50	9	50					Fischer 1970

It will be more useful to discuss breeding success later but it is worth looking in greater detail at the first three eggs analysed. DDE and PCBs were detected in all three but in the greatest concentration in the Outer Hebrides egg. This pair were known to prey heavily on fulmars and some other seabirds, which at that time were accumulating the now-banned agricultural pesticides that were still leaching off the land. The overall levels may not have been high enough to cause infertility, and indeed the embryo in one egg had undergone partial development (Dr Mick Marquiss, pers. comm.). Furthermore, Mick informed me that none of the seven eggshell samples measured at the time was unduly thin, showing a mean thickness of 535 ± 38 μm (range 472–591). By 1987 there was little, then, to implicate pollutants in egg failures, and there is even less nowadays.

The embryonic Scottish breeding population continued to increase in number, albeit still slowly. A total of eleven pairs were on territory in 1988, of which at least six produced clutches. Two other pairs that had bred in the past were not relocated so it is possible that they, too, bred. Sadly, both successful pairs in Mull failed during incubation that year. A new pair – both released in 1982 – which had formed the year before did, however, lay eggs for the first time, at the end of March, and by early August had fledged young. They were not ringed or tagged but were probably both females. This was the only pair to be successful in 1988. Over one quarter of all released birds were now known to be breeding or at least holding territory, and seven of these birds were still bearing tags from Rum.

About a dozen sites were active in 1989 with eggs known to be laid in five nests. Blondie and her mate reared two young. The other Mull pair nearby laid the first of their three eggs on 9 March. Not only was this the first three-egg clutch of the project but it was also the earliest laying date so far; they fledged two young by early July. The only successful pair from the year before hatched a tiny chick by mid-May but it disappeared a few days later. However another pair that had been failing for the past three years at last fledged a single chick, between 22 and 25 August, aged about 14 weeks.

Normally an eaglet leaves the nest at around ten weeks of age, but some may wait longer before taking the first flight, Norway. (Harald Misund)

Among the 18 clutches in our project for which data is available, five had contained a single egg and twelve had contained two eggs; with 1989's three-egg clutch our average reached 1.8 eggs per clutch, similar to that recorded in Norway. Although only five young fledged, that was still our best annual production to date. Five years' data provides a small sample but three of the successful breeding attempts contained a single young, while five others consisted of twins. This average of 1.6 young was again similar to that found in Norwegian nests. Since 1983, however, there had been thirty breeding attempts in all, only eight (27 per cent) proving successful, less than half the reproductive success found in Norwegian Sea Eagles.

Although more territories became active in 1990 and nine clutches were laid, only two young were fledged from two nests. Over the previous five years Blondie and her mate had produced seven fledglings, all but her first attempt being twins; sadly in 1990 they abandoned their eggs by the end of April. The other pair nearby excelled themselves by commencing laying even earlier – on 4 March – although perhaps a clutch of only two eggs that year. One had hatched by 11 or 12 April and the chick fledged by 2 July. The Lewis pair failed again, and below the nest there were at least a dozen fulmars and a shag; the field worker also noted a ledge nearby 'knee deep in fulmar feathers'. The other 1990 chick came from the pair that had reared twins in 1988, having failed in 1987. Most of the prey taken to the nest appeared to be gulls and fulmars but there were many rabbits too.

Although one of the unsuccessful Canna pairs tried again, they had

Stuart Rae and Ken Smith fitting a camera at an eyrie in Canna, 1999

Stuart Rae descending into the Sea Eagle eyrie in Canna

Video recording equipment in weathertight box back from the clifftop, with monitor screen; solar panels out of shot, Canna, 1999

failed by 1 April when Stuart Rae, Mike Pienkowski, Ken Smith from the RSPB and I checked the nest to rig up a video camera. It revealed that the pair continued to return to the nest ledge. At other times ravens also visited to rummage through the nest cup and to steal sticks and this might implicate them in the nest failure. On 3 and 4 April even the neighbouring pair of Sea Eagles landed on the nest ledge. The male of the resident pair, with a white tag in one wing (from 1984), who had ousted the original male the previous year, roosted on the nest ledge overnight. The camera was dismantled on 4 May.

The camera sent a black-and-white frame every few seconds to a weathertight box several hundred yards away, inside which was a video recorder powered by a tractor battery, which was in turn trickle-fed from solar panels. I visited the box every ten days or so to change the video tape. I remember, after several visits, going straight to the spot in a thick mist. Quite proud of myself, I serviced the video but when I set off to return home I found myself totally disorientated and standing right on the edge of the 200-metre cliff top. Such dramas notwithstanding, this self-contained system, perfected by Ken Smith, showed considerable promise so we resolved to monitor nests again next year.

In mid-February 1991, before any eagles were in residence, we visited both Canna and Rum to install four cameras, two of them above nests used the previous year, one at a new nest, and one at a roost site which had always looked like a potential nest. Only the new nest was ultimately used, two eggs being laid around 23 March. One of the units gave somewhat erratic service while a second – at the summit of the 200-metre cliff – often found itself sitting in thick mist, which interfered with the solar panel top-ups. We did, though, get some interesting information.

At nest I, on 8 March, a tagged immature Sea Eagle, probably reared in Mull in 1989 and which I had seen in Canna the year before, landed below the ledge. One of the previous year's occupants at this site, with a tag in its left wing, then landed on the ledge with a fish on 22 March and probably roosted there overnight, and again on 26–27 March. One or two other tagged birds visited but no breeding took place here so, after technical problems, I dismantled the system on 7 April. The pair seemed to break up, confusion over this territory arising because it appeared to extend between Rum and Canna, with occasional interruptions from a third, intruding Sea Eagle.

Nest 2 in Canna was on the cliff top which, with nest-building activity apparent, looked more hopeful. Mist interrupted filming from 16–23 March and it is thought that two eggs were laid during this time. I supplemented the slightly hazy monochrome video tape frames with telescope observations, from below the nest and from some distance away. There were frequent visits by a sub-adult Sea Eagle which once flushed one of the adults from a ledge near the nest, holding a shag in its talons.

Moulting Golden Eagle in flight, Canna

There were also occasional fly-pasts by the neighbouring Golden Eagles that had their own eyrie further along the cliff face, less than a quarter of a mile away. Remarkably there were few aggressive encounters and the goldens successfully fledged their own eaglet, as they did on many occasions in succeeding years, despite their new neighbours. Both Golden and Sea Eagle eyries were north facing along the same stretch of cliff and while the Sea Eagle ledge looked to be a substantial one, I could see the edge of the eyrie structure being tossed around alarmingly in strong gales with fierce updrafts. The nest was finally abandoned on 26 April; it may only have been coincidence but it was during another spell of intense interest shown by the sub-adult. Crows also visited the ledges around the nest to rake through the prey remains. The adults continued to visit the nest occasionally, once to eat a Fulmar kill, until I finally dismantled the set-up on 9 May. Carrying equipment, especially the heavy tractor batteries, up and then down the steep slope to the cliff top was an exhausting undertaking. I never was able to work out from the camera evidence just why it was that these Canna pairs repeatedly failed in their breeding attempts and it would be some years before a Sea Eagle would be reared on the island.

Elsewhere, one of the Mull males had attracted a second female, both his partners laying eggs with him and taking incubation duties at both nests! Remarkably, one of the clutches hatched on 22 April and its twins fledged around 12 July. With the male of the original trio now finally settled with only one of his females, this territory also produced eggs, around 10 March, and fledged two young in early July. One of the young, a male, was found dead about 5 km away on 10 September. In 1991 eight pairs laid clutches, four of them resulting in seven chicks, the highest number to fledge in one season since the project began. One of them would later be seen flying over Dalswinton, in Dumfries and Galloway, on 30 March 1996.

In 1992 the new trio persisted again, resulting in two clutches. The male retained his commitment to incubating at both sites but only one nest produced twins. One of the progeny was later found dead, possibly shot, on 19 February 1993. The other went on to charm Mull visitors and TV audiences alike for years to come and quickly became known as Frisa. The neighbours

A third year immature Sea Eagle tethered in Rum, 1979

(two of the very first trio) produced three eggs for the second time from which two chicks fledged. One of them was found dead beside Loch Morar near Mallaig on 8 February 1993; its carcass contained lead pellets so, if this had not been the ultimate cause of death, the bird had been shot at some stage in its young life. Video cameras were again installed at the Small Isles cliff but, as I had fretted would happen in the previous year, the eggs must have been blown out in gales accompanied by horrendous updrafts. Our growing Sea Eagle population nevertheless repeated the previous year's record success, with seven chicks fledging in 1992.

It seemed to be an appropriate moment to reassess our progress to date. Over the winter, with input from Dr Mike Pienkowski, from me and the project team, Dr Rhys Green – then with the RSPB – analysed the data (Green, Pienkowski and Love 1996). The age of first breeding was known for fourteen established breeders and averaged six years for both males and females. The youngest was a male who bred at three, with his mate – Blondie - four years old; their first two attempts failed, but they were the first of our pairs to produce a chick, in 1985. Laying dates were spread from 4 March to about 16 April, with a mean of 24 March. With the exception of the replacement eggs laid in the second week of May, 50 per cent of clutches were laid in the third and fourth weeks of March. With the welfare of the birds being paramount, nest inspection by the project workers was always tentative, but the size of 22 clutches is known, ranging from 1–3, with a mean of 1.9 ± a standard deviation of 0.42. Over ten seasons (1982–1991), ten or so territorial pairs laid a total of 48 clutches, 14 proving successful and resulting in 22 young fledged in total. However, all of these came from only four pairs and nearly three quarters from only two pairs – Blondie and her mate, and the pair that had been part of the original trio.

Only six clutches were produced in 1993, from which five young fledged, including only one set of twins. In 1994 ten pairs held territories producing eight clutches, although again only five young fledged with one set of twins. However, a new and significant milestone in the project had been passed, for a new pair was formed, the adult and her young mate both having been bred in the wild in Scotland. Their clutch failed to hatch but a new and important phase had already begun.

Chapter 11
Further developments

Sometimes eagles' wings,
Unseen before by Gods or wondering men,
Darkened the place.

John Keats (1795–1821)

By now the membership of the Sea Eagle project team had became a lot more fluid as people moved in and out of other roles within their own respective organisations. Too many names came and went to name them individually but all played their own, invaluable roles. Roy Dennis and I remained the only two founding members from the earliest days of the project, latterly – as we too moved on – acting as independent representatives. Other sub-groups were spawned when necessary, so the original remit became that of a UK project team, still meeting annually, latterly with Richard Evans, former Field Officer in Mull, as Secretary, with the Chair rotating between the RSPB and Scottish Natural Heritage, the successor body to NCC.

As already discussed, Rhys Green had collated the available data to construct a computer model of our population's prospects. He had concerns that 90 per cent of the young reared in the wild so far had been fledged from only three home ranges. Although survival and productivity had been encouragingly high, and was assumed to improve further as time passed, there remained a distinct possibility that the still small population might slowly decline again to extinction within a century. Rapid release of a further 60 juveniles into the population would substantially reduce the risk of this happening. And so, retracing our steps a little in our narrative, Phase II was launched in 1993…

Phase II: Wester Ross

This time, more sophisticated cages – based on a model pioneered by recent Red Kite reintroductions in Britain – were erected overlooking the shores of Loch Maree in Wester Ross. The release site on Letterewe Estate was kindly made available by the owner Paul van Vlissingen and his staff. Here, young Sea Eagles were freed into a wider range of habitats including woodlands and freshwater, yet still within reach of the sea. Kate Thompson, then Kevin Duffy, looked after the birds. Kevin wrote:

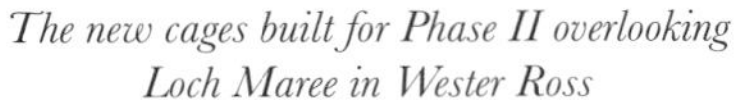

The new cages built for Phase II overlooking Loch Maree in Wester Ross

Harald Misund, Greg Mudge and Kevin Duffy at the back of the Phase II cages, Wester Ross

It doesn't get much better for a bird conservationist. For four summers I was lucky enough to rear and release Sea Eagles for SNH in a spectacular part of Scotland. Based in a bothy with no electricity, it had one piece of technology – a phone. I did have a heated shower powered by gas bottles and a couple of rings to cook on, but other than that just an open fire. Pine marten, wildcat and badger outside and the previous year's released Sea Eagles hanging around too, and sometimes visible even from the window.

Sounds great, and it was. But, of course, nothing is ever that idyllic, is it? The site did host the most voracious midges in Scotland, as well as ticks and some biting things I'd never heard of. If you were lucky you could get bitten by five species of invertebrate in one day! Not to mention the rain . . .

Overlooking the Bodø coastline from Falkefjellet summit, Norway (Harald Misund)

The author back in Nordland to collect Sea Eagles for Phase II (Harald Misund)

One of the chicks destined for Scotland

Once again the Sea Eagle chicks were collected by Harald Misund from the Bodø area in northern Norway. I was given leave from my new post as SNH Area Officer for Uist, Barra and St. Kilda to fly out to help Harald from 2–7 June that first year (1993). Throughout Phase II we were again privileged, through the good offices of Flight Lieutenant Steve Rooke, to have the continued co-operation of 120 Squadron, RAF Kinloss. Tragically, Steve Rooke died, aged only 48, on 25 January 2002.

After all ten birds had been ringed, tagged and released over Loch Maree they remained in the vicinity until at least the end of November. During this time they did range as far as 40 km away but returned to feed at the dump and to roost in nearby trees. In mid-December several moved on (to Rum, Harris, Glen Affric, Inverness, Golspie and Aberdeenshire, on the east coast, for instance) but three or four remained around the food dump until mid-March 1994. By this time, too, at least one bird was being seen in Orkney and Shetland.

Close-up of Sea Eagle chick collected in Norway

Installed in its new 'nest' in Scotland

Posing for the camera, Wester Ross

Sharing its nest platform with cage mates, Wester Ross

Feeding time, Wester Ross

Just noticing the camera through the peephole, Wester Ross

Not pleased at the intrusion! Wester Ross

On 28 June 1994 ten more birds were flown in from Norway. As before, their staple diet was fish, supplemented by rabbits and other miscellaneous carrion. An average of 1.2 kg of food was provided per bird per day for the 35-day quarantine period. All ten chicks were successfully released in early August, slowly dispersing from the food dump over the next few months. Several turned up on the east coast, especially at Munlochy.

One of the 1993 releases (Red 7) had been last seen at the release site on 4 November that year, turning up at Munlochy Bay near Inverness a month later, where she remained until early February. The next sightings were on Orkney on 24 and 25 April. The very next day she was at Sumburgh in Shetland, but returned to Orkney a few days later. Then there was nothing until, on 29 October 1994, a bird with a tag in its right wing, whose colour combination matched Red 7, was seen perched on a rock on the shores of Fraena, an island in west Norway! This was the very first instance we had recorded of one of the reintroduced eagles returning to its native land.

Released in Scotland in August 1993, Red 7 visited Munlochy, Orkney and Shetland before returning to Norway, first Fraena by 1994, then Rogaland, where in 2002 she bred successfully for the first time (Photographer unknown)

Red 7 was identified again in May 1997, further south in the district of Rogaland. In 2001, in her ninth year, she paired with an unmarked male and although they built a nest in a pine tree, she did not lay. She fledged her first chick the following season, however, with twins in 2003, a single in 2004 and another in 2005. Unfortunately for us, Red 7 lost her one remaining wing tag during 2005 so is now unmarked and we have lost track of her. It was probably fortunately for her, though, since the tag looked so ugly and may even have been an encumbrance for her. By this point, I was keen to have the tagging programme stopped; we had gained some fascinating insights but I no longer wanted our wild Sea Eagle population in Scotland to be sporting such obtrusive bits of plastic.

Roy Dennis had always suspected that some of his Fair Isle releases from 1968 might have made it back to Norway which, on a very clear day and with sufficient altitude, they might have been able to see from Shetland. This incident now gives some credence to his idea. We had always been of the opinion that the reverse would never happen, Shetland being too small and low-lying to tempt a Sea Eagle to strike out across the North Sea against the prevailing winds. However, an immature Sea Eagle, colour-ringed as a chick in Norway, did just that, and was spotted in Shetland in June 2001. Then, in March 2004, a Finnish colour-ringed individual was sighted at Loch of Strathbeg, a few miles south of Fraserburgh in northeast Scotland. Most recently a young Sea Eagle colour-ringed in Norway in 2011 was sighted in Unst, Shetland, during December 2012. Such instances are so rare that our population in Scotland are unlikely ever to re-establish by such a route.

Poor weather in Norway in 1995 made the Phase II chick collection rather difficult for Harald, so he only succeeded in providing six chicks for Scotland. They were released from

A tree eyrie at Festvag, Mistfjorden, Norway (Harald Misund)

the Letterewe site during August, all six remaining around the food dump until early October. From sight records at least five 1993 releases could still be accounted for up to the end of 1996. At least seven 1994 birds were known still to be alive in October 1995, and six of them until December 1996. The seventh had been seen at Gairloch, near the release site, between March and May 1995, sometimes with up to four other Sea Eagles, but were not sighted again until 18 and 19 September of that same year, around Loch Teacuis in Morvern, West Inverness. In July 1996, however, it was found dead by a hill walker at Langwell Water, Berriedale, Sutherland; the bird had been poisoned.

Three of the 1995 releases provided interesting sightings. A sheep farmer watched one attacking a fox near Dingwall on 12 February 1996. During the spring, too, another briefly joined a falconry centre's Bald Eagle being flown from the fist near Beauly. A Sea Eagle was even seen flying over the centre of Glasgow on 18 March 1996.

On 16 June 1996 I flew out to join Harald once again and the next day was helping to feed the eagles he had collected so far. One was from a brood of three in Kjerringøy. Harald was trying to select smaller birds in the hope of redressing a slight imbalance towards females. After cleaning up the eaglets' mess in his cellar and driving to the town dump with the guano-ridden debris, Harald took me up to his cabin in Falkeflaug:

> Harald told me there are lynx, wolverine and hawk owls. He only remembers two lemming years in all his time there, when the rodents were in abundance and even swimming on the loch. He also pointed out a small clump of birches where bluethroats breed. The area used to be occupied by Saami, before being bought by a Norwegian at the end of last century, who then sold it to the grandfather of Harald's wife Marianne. Two families lived there at the time and there was even a small school, but Marianne went to senior school down by the coast. She had seven brothers who used to spend the winter at the coast working as masons, building piers etc for the cod fishery, or as wheelwrights. In the summer they returned to Falkeflaug to help with the farmwork. They cut hay to feed the calves, some of which were bought in to fatten, and the good grazing meant that they fetched good prices. They also grew their own vegetables, fished, gathered multibaer (cloudberry), fished trout from the lake and hunted grouse/elk.

Harald and I at his cabin, Falkeflaug

At the turn of the century there was a bear at the head of the valley. In 1987 I had gone just above the treeline to see the tree where it had once clawed the bark. The local Gyr Falcon was not breeding that year. Marianne's family used to carry all their provisions on a cart with horses; the track and bridges round the lake are still in good repair (but do get washed out some winters). When Marianne's grandmother was married her brothers built a resting chair into the rock by the track, and carved 'Bride's Chair' (Braidstole?) into it. The family moved to the Bodø area, where Marianne's father and aunt still lived when I first went there. Her brothers maintain the old family farmhouse as their cabin, while Harald built his own nearby. An old man called Fjelsa had a cabin further up and last winter (1995–1996) when aged 85 he went up to live in it for two months! His son Jon Fjelsa is the renowned bird artist and ornithologist (specialising in grebes) based in Denmark. Marianne's family made some money selling fishing permits and have to pay 125 kroner to the State for each elk they shoot. One taken each winter is sufficient for Harald's family needs.

That evening Harald took me into Gildeskal, an area I had not visited before. Here we found a spot to overlook the island of Fugløya where the locals used to harvest Puffins; Harald pointed out a Peregrine site and two Sea Eagle sites on the island. We finally located our own Sea Eagle nest nearby where we were able to take one chick from the brood of two. We had our quota of ten birds by the next day. Then Harald was able to accompany me in the Nimrod back to Scotland. As we struck out across the North Sea Harald looked down on his home town of Ålesund:

> This time I have a fine chance to accompany the birds in the RAF Nimrod. I had first set my eyes on this beautiful aircraft (then called the Comet) when I went to Gibraltar by boat in 1953. The smell of a British plane with fumes from ten young eagles is one to be remembered! We followed the coast of Nordland… Below was the bird island of Runde, while on the starboard side was Godøya where, as a boy, I saw my first White-tailed Sea Eagle. Now I am glad that children in Scotland can have a similar experience. But perhaps best of all for me are all the positive and helpful people I have encountered along the way, both in Norway and in Scotland. Landowners around Bodø are proud of their sea eagles and were sometimes reluctant to see them disturbed.

All ten eaglets were released by Loch Maree and, along with two of last year's releases, freely utilised the food dump. In June a year later, one was spotted in Skye feeding on a freshly-caught Guillemot. On 27 June 1997 ten more birds arrived in Scotland, this time flown to Kinloss courtesy of the Norwegian Air Force as a gesture of support for the project. The eagles were released during August and continued to feed at the dump into October, while further sightings from previous releases came in from all over the Highlands and Islands – Carrbridge, Easter and Wester Ross, Sutherland, Skye and Mull.

1998 was the sixth and final year of the Phase II imports and once again the RAF undertook the transfer of twelve eaglets back to Scotland. Magnus Magnusson, then the Chairman of SNH, was there to welcome them. The birds were released at Loch Maree during August but by the 31st one was found to be in poor shape and had to be recaptured and put into veterinary care, with little hope of it ever being released. Unusually, the others chose to remain around the food dump for most of the winter, with six still present into February 1999, not dispersing until late March. One was found freshly dead near Lairg, Sutherland on 26 October 1999, seemingly having died of natural causes, probably relating to acute bacterial septicaemia. Another left the release site on 11 February 1999, to be spotted a short distance to the south on 23 April before turning up in Caithness on 13 October. It then moved south all the way to Morven near Mull by 9 December 1999 and was still there on 20 May 2000.

Over the six years of Phase II, from 1993 to 1998, no fewer than 58 more sea eagle fledglings – 25 males and 28 females – were set free. We will now backtrack a little to catch up on our story.

By 1994 ten territories had been established in the west of Scotland, and the same again in 1995. This included not only the trio that had been established for the last few years in Mull but a new trio, again of a male with two females, on the nearby mainland. Here there had been an indication for a season or two that something was about to happen. Eggs were laid in one of the nests but the chick died. The Scottish-bred duo that had established during the previous year failed again. On the other hand another pair (in Canna) that had failed for many years finally succeeded and raised one young.

A new pair became established in Uist in the Outer Herbrides and was of particular interest to me because it was found by my friend, Archie Macdonald, only a few miles from where I live. It had looked promising the year before and although it was not to succeed in 1995, it would become one of our most successful pairs. Depending on which nest ledge the pair chose, we could sometimes sit in Archie's front room with a dram in our hand, watching the pair, albeit some distance away, through his telescope. Now that is my type of bird watching...and an experience I would be able to repeat a decade later, in front of the pub in Shieldaig!

A new pair establish in Outer Hebrides near my home, and destined to be one of our best breeding pairs

This now brought the number of known successful pairs in Scotland to six. In all, nine clutches were laid, one more than in the previous year, and a record six clutches hatched. Five broods survived to fledging, which was also the highest number so far, producing seven young including two sets of twins, thus equalling the record production of 1991 and 1992.

The following season (1996) saw the pair that had both been born in the wild in Scotland rear their first young. My local pair also produced a chick, meaning that we now had eight Scottish pairs capable of rearing young. In all, a record twelve clutches were laid and a record eight clutches hatched. Seven broods survived to fledging, which was also the highest number to date, with a record nine young fledged.

Archie Macdonald and the author ringing the first of our local pair's many chicks (Mairi Macdonald)

In 1997 Blondie and her mate brought off twins. The first original trio (now just a pair, of course) reared a remarkable brood of three. My pair, however, with a north-facing nest, failed, we think during a severe gale. A new pair consisting of a 1992 fledgling and a three-year-old Scottish-bred male also failed at the chick stage. An extra few pairs had now established and we think, from 11–13 clutches laid, six hatched and five of the broods produced nine chicks, including our first ever triplets, equalling the record of the year before.

To this point, progress had seemed agonisingly slow. But by the time Phase II ended in 1998, two established pairs now included some of its first releases (two birds that were still only 4 years old and one of 5); satisfyingly, they both bred successfully, producing three young between them. That season also saw our population now standing at 18 pairs which produced 16 or 17 clutches and nine broods totalling 13 young – a new record production. One of the Scottish-bred pairs fledged a chick that time and the Small Isles trio produced another chick. My home pair produced twins – the first of many to come. Two pairs failed, however, due to the grievous actions of egg collectors; the pair that had reared triplets the previous year were robbed only days from hatching, while the other pair had also had an excellent breeding record up until then.

In 1999 our population still stood at 18 pairs but they produced only eleven young, two fewer than the previous year's record of thirteen. Strangely, an apparently healthy 3½-week-old chick was blown out of the nest in a gale. One of the last chicks imported from Norway the year before was found dead in Sutherland; a post mortem revealed no suspicious circumstances. It was only the third of the Phase II imports to have been recovered dead thus far.

A juvenile male that had fledged in the wild in 1998 – Green E – was found badly injured in December that year. A vet successfully operated on its broken leg, enabling the bird to be released back into the wild on 11 February 1999. It was later seen in Canna, on 24 March 2000. Sadly, Bob McMillan found the adult Green E dead in a forestry plantation in 2003, perhaps as a consequence of a territorial dispute, but its body was too badly decomposed to reveal anything useful. Another adult, apparently showing evidence of a fatal heart disorder, was found under a wind turbine in Skye in 2000, but no link could be proved.

Over the years birds at Site 17 have used several different nest sites all close to one another

Interestingly, in June 2000, an adult Sea Eagle was seen following a fishing boat off Turnberry in Ayrshire, perhaps the same bird that would be seen off Helensburgh exactly a year later. It did not breed but it was unusual for adult birds to be wandering around like this and demonstrated just how much the Sea Eagles might be spreading their wings. By now Rhys Green had re-run his computer model and was encouraged to find an improved breeding success in our population, with the long-term survival looking much more assured.

The year 2000 marked the 25th anniversary of the Rum reintroduction with a significant total of 100 young now fledged in the wild. Appropriately, 25–26 territories had been established, although in this landmark year only 21 of them were still occupied. Three of these pairs were breeding for the first time. One of them consisted of a three year-old male (released in 1997) and a four-year-old female (released in 1996) which – remarkably, considering their immaturity – fledged a single chick. Three other pairs, all derived from Phase II releases, bred successfully. Thirteen chicks fledged altogether. Thirteen pairs had now demonstrated an ability to fledge young. One of these (pair 8) had produced a staggering 21 young in 13 breeding attempts, but this and three other pairs, including Blondie and her mate, were still responsible for 68 per cent of the Scottish-bred young. Since these territories had been occupied by the same individuals all this time, we were becoming slightly concerned by their advancing age.

And sadly, for this very reason, the year 2000 was to see the end of an era. Dave Sexton's pin-up, Blondie, died. He has penned an eloquent – indeed emotional – tribute:

In April 2000, I'd watched her at the nest. She'd just hatched her latest chicks and was brooding them carefully – as she had always done. As I turned away to continue my walk, I couldn't know it would be the last time I'd ever see her. A week later I was called by Richard Evans, the RSPB Mull Officer at that time, to say that Blondie was missing, presumed dead. He thought I should know.

She simply vanished one day in April. It was and remains a mystery. She was too good and attentive a mother to abandon her new brood. Something else must have happened, perhaps a territorial clash with another eagle? We will never know but sadly we never saw Blondie again. Despite our best efforts and a helping hand with food, her bewildered mate just couldn't raise the chicks on his own and, during a spell of wet, cold spring weather, they too perished. For weeks, then months he waited for his mate's return. They'd been together for over 16 years. We'd find him perched at many of their favourite old haunts, his head occasionally tilting upwards to eye a high soaring eagle, to wonder and hope. But his wait was in vain.

After a few days his lonely vigil ended and he eventually drifted away from the nest to spend his days searching for Blondie. He would often sit nearby and just call and call – but there was to be no answer. She was found many weeks later, high on the hill, lying in the heather. It was too late to carry out tests to determine what had happened to her on that fateful hunting trip but we suspect she'd been the target of an attack by another eagle – perhaps a neighbouring Sea Eagle or Golden Eagle whose territory she had accidentally strayed into.

I'm pleased to report that one year he did eventually find a new partner and they bred successfully for another three years before he too vanished from the hills and glens forever. That's the way of it. But the territory lives on with new occupants. And their offspring live on all down the west coast of Scotland. Blondie's dynasty is epic.

In 1978 she had been taken from her nest in Norway as a fledgling as one of Phase I's reintroduction project. Flown across the North Sea by the RAF she was released in the Isle of Rum in Scotland a few weeks later, but quickly found her way to Mull which would become her home for the rest of her long life. Here she paired up and for the next 18 years she would become mother to 17 young Sea Eagles, including seven sets of twins, some of which are still with us and breeding successfully today. She was a true pioneer in every sense of the word and it is no exaggeration to say that without her and her mate's productivity, the whole Sea Eagle reintroduction project would not be the success it is today.

Her chick from 1992 is none other than our very own Frisa (star of BBC's *Springwatch*) who will be 20 herself in 2012. Only this week I was watching the fabulous Frisa with the winter's sun beaming down on her and thinking how much like her mother she looked. Of course to some, all White-tailed Eagles will look the

A celebratory event held on 30 November 2000 in the Aros Centre, Portree, Skye. The front row, from left to right, is Stuart Housden (RSPB), Roy Dennis, the Norwegian Consul in Scotland, Harald Misund, the author and Mike Scott (SNH); behind are friends from South Uist and Skye (Paul Boyer)

> same. Maybe they do? But for a while there I was transported back 25 years to those lonely but wonderful days on the edge of the loch watching Frisa's mum Blondie raise the first wild-bred chick in the UK for 70 years. How proud she would be of what Frisa and her faithful mate Skye have achieved since they first paired up in 1997 and all the chicks they've raised to fledging over the last 12 years. Blondie was just a stunner in all respects: always immaculate plumage, devoted mother and faithful mate and she's why I do what I do for the RSPB to this day. At least I can still watch Frisa today and sometimes catch a glimpse of Blondie in those piercing sunlit eyes.

Eventually, as Dave says, Blondie's widow found another mate and would breed for another three years before he too disappeared. Frisa, one of Blondie's daughters, did become 20 in 2012 and, with her mate Skye, successfully fledged yet another eaglet.

A celebratory event was held at the Aros Centre in Portree on 30 November 2000. Members of the Sea Eagle project team were joined by over 120 guests, including crofters, landowners, foresters, councillors, local community groups, such as the Mull and Iona Community Trust, and tourism organisations. Roy Dennis and I each gave a talk before the screening of some black-and-white film from Fair Isle in 1968. Harald Misund had flown over from Norway to attend, along with the Norwegian Consul based in Edinburgh. Harald spoke movingly of how his 40 years with the Norwegian Royal Air Force had meant little to him compared to his work with the Scottish reintroduction. The folk musician and composer Alan Reid (formerly of Battlefield Band) performed his song Iolair na Mara from his first solo CD, entitled *The Sunlit Eye*. Alan told me how he had heard me on a radio programme about the Sea Eagle project and my mention of 'the eagle with the sunlit eye' stuck in his mind, eventually to give him inspiration for the song.

Also in the year 2000 an international conference on Sea Eagles was held in Björkö in Sweden, where we were able to compare notes with studies elsewhere on the Continent. A symposium volume was published in 2003 (Helander, Marquiss and Bowerman 2003).

In 2001, for the first time, Sea Eagles from Phase I were outnumbered in the breeding population by wild-bred birds. From 23 occupied territories, seventeen pairs were known to lay eggs, including two new pairs consisting of a wild-bred bird mated to a Phase II release, the first time this had been noted. In all, only seven pairs were successful, fledging eleven young. Nevertheless, the remaining ten pairs had failed, albeit due, it was thought, to natural causes, mostly during incubation. Blondie's widow, with a new mate, produced eggs but their attempt failed soon after hatching. Even Pair 2, the former first trio, failed, for the first time due to natural causes since 1988.

Three new territories were established in 2002 with 22 pairs laying eggs but with eight pairs failing during incubation and other failures in the early chick stage, only eight broods fledged successfully, bringing off a total of 12 young. One of the failures was the result of the poisoning of an adult male at Loch Morar. Although both its chicks were fostered into other nests – the first time this had been achieved in Scotland – only one survived to fledging. The orphan later seen in Mull, with tag Blue 1, had been called Dalta – a Gaelic word meaning 'foster child' – by primary school children in Skye. It was later seen at Shieldaig at the end of September 2005. Alison MacLennan of the RSPB concluded:

> The survival of this chick to fledging was a massive achievement. Everyone who has been involved with Dalta is overjoyed to hear that she has been spotted. It is quite incredible that even though sea eagles are our biggest bird of prey in Scotland, they can disappear for such a length of time and then turn up out of the blue like this.

A montage by the author of the delegates at the international Sea Eagle 2000, Björkö, Sweden; the organiser, Bjørn Helander, is in the centre

Loch Morar, Inverness-shire

The orphaned White S from Loch Morar being tagged just before it fledged (Martin Carty)

As usual, several hundred sightings were reported in 2002, but none were more comprehensive than those of two young fledged the previous year. White S had been reared in 2001 by the male that was to be poisoned the year after at Loch Morar. In January 2002 a male White S was seen on Benbecula, then South and North Uist during March. By the end of the month he was sighted at Orbost in Skye and four weeks later in Moidart, West Inverness. A fortnight later he turned up in Caithness but returned first to Canna on 27 May and then to Skye by 4 July. 23 September saw him in Rum, he was back in Skye in July 2003 and then in Lewis on 6 December 2003. He seemed to have paired and settled in Skye in 2004, but by 2011 he had not met with any success and seems to have disappeared in 2012. Quite a remarkable odyssey, though. One hopes that he may yet turn up again…

The other tagged immature, White T, had fledged at Orbost in Skye in 2001 and proved just as adventurous. In January 2002 he was in Canna but in Sutherland and Easter Ross by the end of the month. On 8 March White T was in Wester Ross, in Mull a week later and Morar by the end of the month. In mid-April he visited Glenbrittle, near his natal territory, but came back to Mull six weeks later, then was in Argyll in August and Wester Ross in November. White T was seen on the mainland beside Mull in February 2003 but was back in Gairloch, Wester Ross by July. In March 2004 he was interacting with an established breeding pair although visiting Lewis four months later. He finally ousted the older male the following year but it would be another two years before their breeding attempts proved successful, as they have in most years since.

2003 was to prove a better year as our population continued its slow increase to 31 territories. Dave Sexton gave up his post at RSPB headquarters to return to Mull, where he has remained ever since. From 25 clutches laid in 2003, 20 hatched but, despite losses of some very young chicks, 16 pairs reared broods, twice as many as in the previous year. A record 26 young Sea Eagles fledged. The nine failed pairs all seemed to be victims of natural events, as were two young adults found dead in 2003, one of them Green E that had fledged in 1998 before successfully recovering from a broken leg. The fate of a breeding female was more sinister, though; the widowed mate of the Sea Eagle that had also died from poisoning the previous year, she was found poisoned in Morar in 2003. She had been able to find a new partner but, alone, he was unable to care for the twins, her fourth brood, in their nest.

Dave Sexton in Mull (Debbie Thorne/RSPB)

I wrote an article for the press entitled *A Sorry Saga of Special Sea Eagles* (Love 2005):

The tagged female poisoned at Loch Morar (Martin Carty)

> Having shown so much support and given so much assistance, it is not surprising that local folk were horrified when a nesting sea eagle was found poisoned in the Morar area in spring 2002. Even worse was the fate of his mate a year later, also poisoned. Not only is such activity highly illegal but it is totally irresponsible, for such lethal baits could be found by any creature, someone's dog or even worse a young child. The police have yet to pin down the culprit but have the co-operation of many local people who remain vigilant to prevent any such tragedy happening again.
>
> The loss of any nesting sea eagles to such vandalism is to be regretted but this particular pair were rather special in several respects. The female had been a Norwegian bird released in Wester Ross in 1996 and her mate, another imported bird, was set free there a year later. They came together at only four and three years old in the spring of 2000, not just to build an eyrie and lay eggs but even to rear a single young. They were the youngest pair ever to nest successfully, on their first attempt, in the history of the project.
>
> It is understandable that their inexperience should have resulted in their chick not taking flight until 29 August, aged 16½ weeks – the latest recorded fledging date. The event was especially notable since this was the hundredth chick to be fledged in the wild. Futhermore, it came from the 25th territory to be established and the event coincided with the 25th anniversary of the Reintroduction Project.

Flushed with their success, pair 25 went on to fledge another chick in 2001, but three weeks earlier this time. This bird was fitted with white wing tags bearing the letter S in black, which enabled observers and the public to track its movements for many months (see above).

Pair 25 looked set to fledge twins in 2002 – their third breeding year. But on 8 May the male was found poisoned. Conscious that the female could not rear the eaglets on her own, the tiny 8-day old eaglets were taken from the nest under licence and kept overnight by Alison MacLennan of the RSPB in Skye. One was found a home in another sea eagle nest with a single slightly older chick. The parents were supplied with food at a dump nearby, and although their own chick fledged successfully the little foster chick did not make it.

The other orphan was put in a nest with twins, but, despite the parents being given extra provisions, it did not thrive. So on 10 June it was moved to yet another eyrie containing a single chick nearer its own age. Again the new foster parents were given a helping hand from a food dump established nearby and both chicks fledged successfully in early August. The orphaned eaglet was spotted in Rum some time later, so seems to be doing well. Its foster sibling would later turn up on Skye, Mull and Knoydart.

Although such fostering techniques had been used in other countries this was the first time that it had been attempted in Scotland. The successful outcome meant that something was salvaged from a bleak situation. Later in the year it was obvious that the widowed female had found a new mate and they may well have gone on to breed. But on 26 February 2003 she too was found dead, having suffered the same horrible death as her mate the year before – poisoned. The culprit is widely condemned by local people, who have continued to monitor the site. Although they have seen a couple of young eagles in the vicinity the territory was not used in 2004.

It is sad to lose any Sea Eagle to the project, but particularly so when both were members of a successful breeding pair and had been destroyed illegally in such a vile manner. From the outset this pair began to break records but were never allowed to fulfill their full potential. An awful lot of people, both within the Project Team and in the wider community, are intent upon seeing the Sea Eagle safely reinstated to its former haunts and now must lament the passing of a very special pair of birds. But perhaps the intrepid explorer White S and one of its orphaned younger siblings will continue their proud legacy.

These illegal, barbaric and highly damaging incidents caused outrage locally and beyond. The police and local community are now increasingly vigilant. Both dead birds had been imported from Norway in the mid 1990s and had been breeding together since 2000.

Meanwhile the RSPB had been filming the Sea Eagle story for three years and *The Eagle Odyssey* was premiered at Eden Court Theatre in Inverness in May before touring round many venues in Scotland. The most successful and acclaimed film the RSPB has ever made, it would win at least nine industry awards, not least Best Script at Wildscreen, the most prestigious wildlife festival in the world, in 2004. It is now available on DVD. Cameraman Gordon Buchanan, himself from Mull, would then film his epic *Eagle Island*, documenting the wildlife of his native isle and its eagles in particular.

Southwest Ireland

We have mentioned how both Sea Eagles and Golden Eagles became extinct in Ireland around 1900. A project had begun in 2001 to reintroduce golden eagles from Scotland to Glenveagh in Co. Donegal. The first chick fledged in 2007, with twins in 2009 and three

Killarney National Park, Kerry

eaglets in 2010 and by then, the Golden Eagle Trust had also turned its attention to Sea Eagles. Interest had already been shown some years previously to release some in Inishvickillaun in the Blasket Islands so this time it was decided (with advice from Roy Dennis) to initiate once more the species' return to southwest Ireland, although this time to Killarney National Park in Co. Kerry. Dr Allan Mee was appointed to supervise the project, with co-operation from the National Parks and Wildlife Service of the Irish Department of Environment (now the Department of Arts, Heritage and the Gaeltacht) and the various Norwegian authorities. Allan travelled to join Torgeir Nigård and his team from Projekt Havørn, the Norwegian Institute for Nature Research (NINA) and the Direktorat for Nature Management, to collect the first Sea Eagle chicks. Their licence stipulated that only one chick be taken from nests with twins but twins are not so common in the area around Trondheim, the breeding success sometimes impacted by factors such as late snow. Nonetheless, fifteen chicks were collected in 2007 and flown to Kerry airport, to be met by a deputation of Irish farmers protesting against the reintroduction project because of what it might mean for their lambs. Surprisingly, the use of poisoned meat baits for crows was not banned in Ireland until 2008, while poisoning foxes became illegal there only in October 2010.

Despite another poor season in 2008, 20 chicks were collected in Norway. Twenty more would be imported in 2009, 22 in 2010 and 23 in 2011. Essentially the same techniques were employed in Kerry as in Scotland, all the birds being fitted with wing tags and radio transmitters or, in the case of ten individuals, with satellite transmitters. From the outset, one of the 2007 females spurned the food dumps and spent the following summer in Northern Ireland before returning in the autumn. Another set out to sea from the Dingle Peninsula and reached the Skelligs, 36 km away, much to the consternation of the resident Peregrines and 50,000 Gannets! The next day it was seen being mobbed by Choughs on the Blaskets. It spent the following summer in Northern Ireland before finally returning to Kerry. Most, however, would winter in the vicinity of the release site, some slowly dispersing thereafter. A few moved east along the south of Ireland while others edged north as far as Antrim in Northern Ireland.

From Antrim, of course, it is but a short hop – 20 km – to Scotland and in early February 2009 the first Irish Sea Eagle (a female released in 2007) was seen feeding on a fish by a loch in Glen Garry, but it had returned to Kerry by May. It is now part of a potential breeding pair in Killarney National Park. In all, five other Irish birds were to visit Scotland. One spent a summer in the Cairngorms, 700 km to the northeast. On 29 October 2010, another was spotted in trouble on the north Aberdeenshire coast and had to be rescued. A kayaker found it in a cove near Pennan so Rhian Evans of the East Scotland Project put together a brave rescue. Grease was found to be matting its feathers and seemed to be inhibiting its flight, but it otherwise seemed quite perky, and after veterinary treatment by the SSPCA it was released from the Fife Phase III cages on 13 December. In February 2012 it turned up at Drumpellier Country Park in Lanark, seemingly on its way back to Ireland. On 14 September 2012 it was relocated in the hills above Glencar, Co. Kerry, some 20 km west of its release site, having indeed made its way back from Scotland. Sadly his badly decomposed body was found in Glencar in December; no cause of death could be determined.

Rhian Evans with the Irish Sea Eagle she bravely rescued from the cliffs of Aberdeenshire (ESSE)

The satellite tag on male L, though, released in 2008, told a more amazing story. He left Kerry on 20 April 2009, covering the 260 km northeast into Sligo in a single day before settling in south Donegal. On 21 May he left Antrim for Scotland, roosting on the night of the 22nd in Glen Shiel, and then, on 23 May, flew to Loch Maree via Loch Carron! Moving through Durness on the north coast he first visited Caithness and then crossed to Hoy in Orkney on 28 May, before heading to Rousay and Westray, 1,000 km from Kerry and only 450 km from Norway, so two-thirds of the way home! Three days later, though, he returned south to Hoy before spending the night of 1 June by Dunnet Bay in Caithness. Retracing his route, he then spent the next four months in the Kyle of Durness. Late in September he moved overland to Ardgay on the Dornoch Firth. On 15 October he was in Glen Cannich and then crossed the hills to Kyle of Lochalsh the next day. By 19 October he had turned up in Mull, where he met up with the locals, two of them also wearing satellite tags. He was also seen on Jura and Islay before heading home to Ireland on 11 November, turning up in Donegal two days later. After eight months on the move he had returned to Killarney in time for Christmas. His was an astonishing odyssey, easily upstaging our own Scottish White S!

Dr Allan Mee of the Golden Eagle Trust with two Sea Eagles poisoned in southwest Ireland (Golden Eagle Trust)

Three months after her release, the 2007 bird wearing the first satellite tag was found dead a short distance away, apparently poisoned with Alphachloralose, which to this day is still legally available in Ireland for use against rats and mice. On 18 February 2008 two more poisoning victims were recovered some 15 km from the release site, with a third later in May and a fourth 2007 release also suspected of being a poisoning victim. This was an ominous turn of events and by June 2012, with a hundred young Sea Eagles released in Killarney National Park, no fewer than 22 have been found dead. At least one had been shot; it was found floating in Lough Neagh and, although only its tags and transmitter were retrieved, they showed signs of having been shot through. Two further eagles have been killed by wind turbines. Two found most recently, in May 2012, were also suspected of having been been poisoned, although one of them also had lead shot in his body. Of ten released Sea Eagles which were wearing satellite tags and whose corpses were able to be located, five (50 per cent) had been poisoned. As a result, the Norwegian government hesitated about further exports to Ireland but fortunately have decided to carry on.

If nothing else, such tragic deaths might have helped facilitate the final banning of poisoning (except for rats and mice) in Ireland by October 2010. Sadly, though, it still continues in Scotland over half a century after becoming illegal, albeit at a much reduced level. The Irish Minister of the Environment, Heritage and Local Government said:

> These [new] Regulations are to address the poisonings which resulted in the deaths of twelve eagles and other birds of prey earlier this year [2010]. I am very concerned that these poisoning incidents could damage the projects to reintroduce the golden eagle, white-tailed eagle and red kite which are being funded by my Department. Such actions are irresponsible as well as illegal and they give a very negative image of Ireland's farming and tourism sectors, nationally and internationally… The re-introduction of these magnificent eagles and kites into Ireland will further enhance Ireland's environmental reputation for respecting its wildlife and enable us to harness the associated social, cultural and economic benefits. We now call on everyone to respect the law and protect these birds of prey, which are of real economic value to the rural communities in the release areas.

One of the Rum juveniles on her tether

And the good news is that by 2012 four pairs of Sea Eagles had established territories in Counties Clare and Kerry. One pair, barely mature, laid eggs that year near Lough Derg in Clare but, after incubating faithfully for seven weeks, they abandoned the nest, leaving one egg and a dead chick. This is the first breeding pair of Sea Eagles in Ireland for 110 years! As with the early successes of Golden Eagles further north in Donegal, the Irish Sea Eagles look as though they will certainly come good in the very near future. Some farmers back in Donegal even admit that they now welcome eagles, as they have suffered fewer lamb losses to foxes and crows with eagles in the vicinity, especially if the flock is taken in-bye and away from risk during lambing.

Following the sighting of a Norwegian-ringed immature in Shetland in 2001, another colour-ringed Sea Eagle turned up at the Loch of Strathbeg for a few days in early March 2004. Enquiries revealed that it had probably been ringed in Finland three years before. One of the birds that we hoped might reoccupy the ill-fated Loch Morar territory was found dead in Mull at about the same time. So with this territory remaining empty in 2004, our tally increased by only one in 2004, but 28 of these 32 pairs laid eggs, 19 pairs hatched and 15 pairs fledged 19 chicks. One of them would be spotted over Lennoxtown Hospital in Glasgow in March 2005.

Another territory, this time in Mull and occupied since 1984, fell vacant in 2005. The female had apparently moved elsewhere but this pair had only ever fledged one youngster (in 1996). The Loch Morar territory remained unoccupied. So although our tally increased by only one pair again, 28 clutches were laid, 21 of which hatched. Seventeen of these proved successful with 24 young fledged. One of these, a female, flew from Blondie's old territory. Another was seen on Loch Lomond in November, perhaps the same bird that was identified in Antrim in Northern Ireland two weeks later.

A Sea Eagle chick for ringing, Outer Hebrides, 2007

In 2006 six clutches failed but no fewer than 26 hatched, 21 of which fledged a total of 29 young – our most successful breeding season thus far. An adult Sea Eagle was reported in Orkney but still none breed in the Northern Isles, all known territories being in the west of Scotland and the Hebrides. But by 2007 the range had expanded somewhat, both to the north and to the south, with eleven pairs in Skye, four of them producing five young. A further 11 in Mull and the Argyll islands produced ten young from seven successful pairs. Finally, ten pairs were found on the west Highland mainland and the Small Isles, from which six young fledged.

So 2007 proved an exciting year, with six new pairs bringing the total number of occupied territories to 42. Thirty five pairs laid eggs and 31 of them hatched. Twenty four of these produced a record total of 34 fledged young. Ten pairs occupied the Outer Hebrides, eight of them fledging 13 young. Three years later, in June 2010, one of 'my' local offspring, a female tagged in Uist, would be found 'injured and exhausted' in Loch Bhrollum in Harris by some French yachtsmen. It was taken into the care of the RSPB and fed lambs' livers from a butcher in Stornoway to build up its strength ahead of its release. RSPB Scotland officer Martin Scott said that the bird may have been injured in a territorial dispute with another eagle. A 23-year-old adult from Phase I was found dead in 2007, perhaps the very last of the original Norwegian imports remaining in our wild population.

2008 was not quite to continue the upbeat trend, although 28 fledged young was still the third-best annual total and two new pairs were located. Thirty five pairs had laid eggs although only 21 broods hatched. A large proportion of the failures came during a spell of bad weather coinciding with the incubation period in the Outer Hebrides, where only two of the eleven pairs successfully fledged four young. On a positive note, adults and sub-adult Sea Eagles were now settling in new areas in the west, with the promise of further range expansion. Some wild-bred immatures were even seen on the east coast, while releases from an East Scotland Project, initiated the previous year, turned up in both Mull and Skye.

With the Sea Eagle population firmly hefted to the west coast, it had always been Roy Dennis's vision to see some establish in the east, where Ospreys and Red Kites were doing particularly well. The Sea Eagle project team had already asked the Centre for Ecology and Hydrology (with input from SNH, the RSPB and the Forestry Commission) to undertake a feasibility study on suitable locations. (I was slightly surprised that Shetland was not being considered; while records of released Sea Eagles have been regular there, none have yet remained to breed, despite its apparent suitability in so many respects.) The CEH Report was to highlight the Firth of Tay as ideal. And so, in 2007, Phase III had been initiated...

Phase III: East Scotland

Although still a member of the UK Sea Eagle project team, my involvement was minimal and so I have compiled this account mainly from the detailed reports and entertaining blogs of Claire Smith, the Project Officer and her successor Rhian Evans. A partnership between RSPB Scotland, SNH and Forestry Commission Scotland located a suitable release point in a dense spruce thicket in Fife and, with the blessing of the local farmers, 11 substantial aviaries were constructed, overlooked by strategic security cameras. Other than elaborate improvements to the cage design, the approach would be similar to that taken in the two previous phases.

The food offered consisted mainly of venison offcuts, fish – including herring waste from the celebrated Arbroath smokies – rabbits and grey squirrels, from a cull being carried out in Aberdeenshire. On May Day, Claire Smith was taken on by the RSPB and was despatched to Norway to link with the head of Projekt Havørn, Alv Ottar Folkestad, who was to supervise the collection of chicks. It had not been as good a breeding year as expected but, with all the appropriate local, UK and international licences, 15 eaglets were collected in the Møre og Romsdal and Bergen areas and, that first year, were flown into RAF Kinloss by the Norwegian

The cages at the new release site in Fife (Claire Smith)

Air Force. Thereafter, continuing their loyal support for the reintroduction of the Rum Sea Eagles, and Scottish Red Kites, some of the flights would be undertaken by our own Royal Air Force.

Although given a veterinary examination on arrival and sexed by DNA, two of the chicks developed respiratory problems, but were successfully treated by vet Alastair Lawrie. One of them, however, Bird C, then suffered an injury in the cage, breaking its right leg and right wing. The vet was recalled and a month later the fractures had healed. As a result a possible defect in the design of the nest platforms was identified and rectified.

The other eagles were weighed, measured and fitted both with wing tags and radio backpacks. While the birds themselves weighed 4.2 –5.6 kg, these radios were only 70g, with a five-year battery life. They were all released in mid-August, with Bird C to follow three weeks later. Only hours after its release, one of them landed on a nearby roadside dyke, much to the astonishment of motorists and a passing cyclist, but they soon became familiar additions to the local avifauna, utilising the food dumps and having their movements constantly monitored by the radio equipment. Within three weeks Bird F turned up at Stonehaven, later moving even further north to Fraserburgh. The others, though, continued to frequent the release site and often roosted together nearby.

Although two farmyard geese were attacked, there proved to be no substance to a complaint alleging the killing by a Sea Eagle of an unspecified number of peahens, bantams, rabbits and cats, nor to another claiming the loss of fifty mallard released by a gamekeeper. These were either panic reactions or even deliberate attempts to condemn the reintroduction project. On the other hand, the arrival and release of the eagles attracted great interest from local schools and interest groups, TV and press. Over the years Claire, her successor Rhian Evans, Bruce Anderson and the RSPB team would put huge effort into public liaison, talks, museum exhibitions and educational school projects. A coloured information leaflet was widely distributed in March 2008 and a short educational film was produced a year later.

Over 400 sightings of the released Sea Eagles had been reported over that first winter.

A male was reported as having been shot in Glenogil, Angus, in November 2007 while White G from Mull was found poisoned in the same area in May 2008. A long-dead female was also found in November 2007 at Forvie National Nature Reserve in Aberdeenshire. Four radios failed but at least two of these birds were known to be still alive. Over that first winter, one 2007 male turned up on the Isle of May in the Firth of Forth, remaining there for six weeks. Other releases spread west into Perthshire, two arriving at the Blair Drummond safari park, where one of them perched in a tree to observe a captive Sea Eagle below. When Bird F first arrived from Norway he was found to have two white claws. He was the first to be released and immediately set off for Montrose Basin, continuing north to spend three-and-a-half months at the Loch of Strathbeg and St.Fergus near Fraserburgh. He then moved on west to Findhorn Basin and upriver to Grantown before arriving in Mull. He was only found when Claire Smith arrived on the island in early March, armed with her radio tracking equipment. Bird F was seen near the territory of the celebrated Mull pair Frisa and Skye, before returning to the Perth area in July.

Further improvements had been made to the aviaries before they were disinfected ready for the eaglets' arrival. In June 2008, 13 young were imported into Edinburgh Airport, their collection having been hampered by bad weather and another poor breeding season in the Bergen area. Two more arrived into Dyce near Aberdeen a month later. Again, two showed signs of chest infections and were successfully treated as a precaution. All fifteen birds were released in mid-August, one release going out live on BBC Breakfast TV.

They seemed to spend longer near the food dump than had the previous year's birds and, in September, two even shared the handouts being provided for reintroduced Red Kites near Stirling. This would become a fairly regular attraction. Being such sociable birds, Sea Eagles established several communal roost sites (at Loch Leven, Montrose, the Tay estuary,

Isle of May, Firth of Forth

Loch Tay, Loch of Strathbeg, Tentsmuir and Backwater/Lintrathen) which came to be used regularly by eight or more birds, even attracting back any that might have wandered.

One 2008 male, who came to be known as Ralf, turned up on the Isle of May after only ten days of freedom. He was seen eating a Fulmar chick and pouncing on a young Herring Gull, both of which continued to be part of his diet until – after appearing on YouTube – he decided to head north. By 10 October 2008 he had reached the Loch of Strathbeg but in February 2009 returned to Fettercairn, 90 km (60 miles) to the south. This was just a brief stop, as the next month he was back at Strathbeg hunting Wigeon and other ducks, injured geese, the odd Whooper Swan or, when all else failed, coming down to carrion. He had also been seen at Troup Head, further west, at the Meikle Loch and the Ythan Estuary nearer Aberdeen. He was still at Strathbeg in April 2010, when he was seen with an untagged juvenile Sea Eagle.

Two of the 2007 birds were in Mull in December 2008 but one took only three days to cover the 100 miles back to the River Tay. It is astonishing how quickly the Fife birds discovered the long-established Sea Eagle community on Mull and distance seemed little object to them. The radio back packs made it easier to track the released birds, allowing a detailed picture of dispersal and survival to be put together. One of the first females set free in Fife had remained in Mull and in the spring of 2009 seemed to have paired up with a locally-bred bird, one year younger. Three birds in all were to become resident in Mull. And then, by April that year, the reverse was to happen. Having first visited Skye, a three-year-old fledged on Mull by Frisa and Skye turned up in Aberdeenshire, in the same area as one of the new Fife releases.

In April 2009, however, a one-year-old Fife female was hit by a train, with another suffering the same fate five months later. For some time the latter had been seen, and photographed, lifting up to 30 pheasants from a nearby hatchery. Occasional complaints continued to come in but not all of them proved credible. In August 2009 yet another Sea Eagle (bearing a radio tag) was found poisoned in the Glenogil area, the third victim of persecution near there in less than two years.

A third eagle released the previous year, in 2008, died under electric power cables. Scottish Power erected several attractive perches in the vicinity in a laudable attempt to stop this happening again. By this time one of the Sea Eagles was frequenting the Solway Firth where it remained for several months, at least once crossing the border into England. One of the 2007 females moved out of Mull in May to turn up first in North Ronaldsay and then in Fair Isle before taking only 40 minutes to reach the mainland of Shetland 43 km (27 miles) to the north, beating the previous record by 5 minutes set by a previous wandering Sea Eagle. Another was seen over Edinburgh and in June hung around its famous zoo for a short time.

Two more breeding territories were located in 2009, one in Lewis and the other in Wester Ross. This marked a significant extension in range, both in the Outer Hebrides and on the mainland. There were now nine breeding territories here, four in Lochaber/Argyll and five in Wester Ross. Of 46 territories in Scotland, 24 fledged a total of 36 chicks, two more than the previous high recorded in 2007. Overall productivity was good at 0.92 young fledged per breeding pair, yet a further 15 pairs had attempted to breed but failed, either during incubation or around hatching. Where the Outer Hebrides had suffered in the previous year, Skye bore the brunt of the failures in 2009. Two were the result of disturbance during forestry operations but the cause of the others remains unknown.

June 2009 saw another 15 Sea Eagles imported from Norway, two of the females from a nest of triplets. At Edinburgh Airport the reception committee of project partners and media included the Environment Minister Roseanna Cunningham. During the statutory quarantine period, and only a week after being installed in its cage, one of the chicks was found dead on its nest platform, the first to die in captivity. An autopsy revealed the cause of death to be Aspergillosis so the others were treated with antibiotics as a precaution. One was later treated for an eye infection and released on 25 August, several days after the other thirteen. They made good use of the food dump which was topped up regularly near the release site. Sometimes, during September, up to ten of them roosted together in a wood nearby.

In August another Sea Eagle (released in 2008) was found dead in Glenogil, Angus, this time poisoned, and one of the 2009 releases was hit by a train while feeding on carrion lying on the track. In all seven eagles were reported dead during 2009, bringing the known deaths to 12. Three-quarters (75 per cent) of the 44 released thus far were still known to be alive, though – 11 from 2007, 12 from 2008 and 10 of the 2009 releases.

Tragically, in August 2009, Mike Madders, a pioneer in Sea Eagle monitoring in Mull and elsewhere, in dietary studies (Madders and Marquiss 2003; Marquiss, Madders and Carss 2003), windfarm effects and much else besides, was drowned in Loch Maree, along with his young son. It was he who had radioed Dave Sexton with the message 'I think we're both daddies!' back in 1985. Dave reflected how the success of the Sea Eagle reintroduction was a fitting legacy for Mike. 'He helped them get to this point and should be justly proud of his work.' He is sorely missed.

Mike was a founding member of Natural Research Ltd who pioneered so many projects on birds of prey. Amongst them was the use of satellite tags on the Scottish Sea Eagles. The company had teamed up with Forestry Commission Scotland, Scottish Natural Heritage, the RSPB and Roy Dennis's Highland Foundation for Wildlife back in 2008 when the first two chicks (in Mull) were fitted with tiny satellite transmitters in order to follow their movements. Five more (two from Mull and one each from North Uist, Lewis and Skye) were recruited into the study in 2009. Their subsequent movements can be followed on-line.

The technology would also be applied to the East Scotland project and on occasions even Claire herself took to the wing to track her birds. Initially, though, she relied mostly on sightings of wing tags. Over the Easter weekend of 2010 a Sea Eagle released two years previously in Fife was seen over Newcastle Airport, before moving south to Flamborough Head and Hull in Yorkshire. Other sightings, probably of the same bird, came from Blacktoff Sands on the Humber Estuary, the Yorkshire Dales, Leighton Moss (Lancashire) and Hawsewater (Cumbria). A 2009 male was seen apparently catching frogs near Aboyne in Aberdeenshire in April, whilst another turned up in Shetland. Having remained within 50 km (30 miles) of the release site for eight months it suddenly took to long-distance travel. It was seen at Forsinard in Sutherland on 6 April, was harassed by bonxies on Fair Isle on 21 April and was at Sandwick on the Shetland mainland three days later. Having been around Shetland for some time, the 2007 female reached the remote island of Foula before returning later in August to a roost on Mull. In the spring of 2010 a two-year- old even visited the Isle of Rum where it all began.

On 25 June 2010, 19 Sea Eagle chicks were flown into Edinburgh; two were from a set of triplets and all the rest were singles from twins. They were all released without incident during August and within ten days, on 28 August, the smallest male reached the Farne Islands off the Northumberland coast where, much to the delight of visitors, he caught a shag, fulmars and gulls. A couple of weeks later, perhaps having exhausted the media possibilities, he was reported on Lindisfarne before heading off further south. On 30 August another young eagle had reached the Isle of May from Fife Ness, the third Phase III bird to do so. Both had doubtless been helped by the northerly winds, from which a third youngster had sought shelter by squeezing into a hen coop in Fife and then found itself locked in overnight! It was released, none the worse for wear, the next day. On 3 September another was found injured and had to be put down, while a female was electrocuted in Fife. On 11 October the eagle on the Isle of May was found with a broken leg, shattered beyond repair, so it too had to be put down. It may have strayed too close to a pupping grey seal and been crushed.

On 18 September, after a spell of exceptionally bad weather, yet another newly- released eagle was captured in Ardnamurchan, 100 miles to the west. It was 700 g underweight but after a spell in captivity back in Fife had recovered well enough to be released again, in the company of other juveniles still in the vicinity. The 15 survivors dispersed over the winter, the female twin reaching the River Ythan and Strathbeg in Aberdeenshire, leaving her brother near the release site in Fife. Her other brother, meanwhile, had fledged back in Norway.

On 24 June 2011, 16 Sea Eagle chicks were flown into Edinburgh, with two females coming from a set of triplets. One was found to have gapeworm and then developed a respiratory infection, although she responded to treatment. She had to be 'processed' and fitted with tags and radio a fortnight after the others, but all were released during the month of August. Just over a fortnight after freedom one bird reached Pitlochry and another Newbiggin on the Northumberland coast. The latter was one of three birds that were found in poor condition soon after release and after a short period of captivity could be set free again at the release site. By 21 November all six 2011 birds were still present at the food dump. Very wet conditions were a feature of that August and September, meaning that three birds were grounded after release and taken back into captivity for recuperation. Bad weather may even have contributed to a higher mortality, certainly hindering dispersal during the autumn. Later in November, though, Red A had turned up on Arran but none moved north and one of the previous year's females returned from Mull to the release wood, the first two-year-old to do so.

In July 2011, the RSPB won a Heritage Lottery grant which, along with some Leader money, allowed them to continue the Sea Eagle work. Although public access and education was to form the focus of the project, it also provided funds for East Scotland to add a sixth year of releases in a bid to bring the total imports as close as possible to a hundred. With late snows and storms, though, 2012 proved a bad breeding year in Norway and few broods were found to have twins. The licence stipulates that a single chick should be taken only from nests containing twins. A week of searching passed before Alv Ottar Folkestad and his team located a second nest with two young and in the end they succeeded in collecting only six birds – four males and two females – for Scotland. Rhian Evans, who had taken over from Claire Smith as Project Officer in August 2011, went to Norway to help and has provided an entertaining blog of her experiences. In the end, only 80 young sea eagles were collected for the East of Scotland Project and, following the death of one in 2009, only 79 were actually released.

Sadly, six Sea Eagles were found dead in 2011, one from the previous year and the rest from the first few months after release in that year, as might be expected. In any population of wild animals it is the inexperienced juveniles who face the highest mortality, and this also proves to be the case with imported sea eagles. Some learn to hunt and kill for themselves within days of release but most rely at first upon food supplies left for them by the release sites. Juveniles that fail to utilise this convenient source are most at risk. The cause of death is not always apparent and this failure to use the food provided might well have been a contributory factor in seven of the 24 deaths reported during Phase III to date.

The fact that the Fife birds carry radio tags and frequent an area where the density of people is much higher than usual in sea eagle territory has doubtless increased the detection of carcasses. Five died by electrocution, two more by colliding with overhead cables. Five died after collisions with trains, something we had not experienced in the west coast releases of Phases I and II, although it is a cause of death reported from other countries in Europe.

Rhian Evans and Claire Smith with young Sea Eagles for release in Fife (Dean Bricknell)

The unfortunate victims were presumably hit while feeding on carrion along railway lines. Only two – one poisoned, one shot – appeared to have been victims of persecution. Two others had to be put down after being injured in the wild and a final bird had died while still in captivity.

There have been remarkably few complaints of the East Scotland Sea Eagles interacting with livestock and certainly none – Norfolk farmers, please note – involving piglets. Some incidents were unsubstantiated while a few resulted in problems. Only rarely are the birds actually seen killing, their prey young pheasant or red-legged partridges, a few chickens and one or two geese. Happily, some farmers were reluctant to 'press charges', while others took appropriate steps to make sure it did not happen again. The media, of course, are usually tempted to make a meal of such encounters. Perhaps the most celebrated involved a minister near Abernethy in Fife who, in August 2011, disturbed a young Sea Eagle in the act of attacking his geese. One goose was killed and another injured, but the brave minister – who according to the press had been attacked – managed to knock the eagle senseless so that it had to be held in captivity for a week to recover. Discussions with the good Reverend resulted in the netting over of his goose pens and no further incidents were reported.

In 2011, the 2007 Fife female that had paired up with a Mull-bred bird of the same age went as far as to build a substantial nest in a spruce tree on the island. The pair went on to use the same nest in 2012 and fledged a chick. Amongst the prey remains in the nest Justin Grant found a mink cub – a noble effort by the pair in the control of an undesirable alien species! Not far away a female released on the East coast in 2008 paired up with a male called Mara, satellite-tagged in Mull and star in 2008 of BBC's *Springwatch*. Together they incubated an egg, albeit unsuccessfully on this, their first breeding attempt.

Not only was it intriguing to see how quickly Fife birds reached Mull to link up with the resident established birds, but it is interesting that their first breeding attempts should be on Mull. In turn, several young Sea Eagles reared on Mull have been seen on the east coast with some of the Phase III releases, ensuring a better mix of genes for the future. Even more astonishing are the movements of at least five of the Irish releases, not only in adding to the mix but also revealing just how quickly this took place. The ability of immature Sea Eagles to move quickly and easily around the UK and Ireland is indeed remarkable. It is of course intended that releases will be 'hefted' to where they were set free, with Phase III birds beginning to breed and establishing an east coast population within a few years. Many would like to see a similar venture develop in eastern England, while Wales is also looking into the possibilities. Although immature Sea Eagles wander, their tendency to return to the area in which they were set free means that it is likely to be some time before England and Wales will attract colonists from the population in the north and west.

In 2010 the number of Sea Eagle territories on Scotland's west coast had passed the half century. At least 45 of these 51 territories produced eggs, and from them some 45 chicks fledged. One was reared by a Sea Eagle pair in Skye which had just embarked on its very first breeding attempt. They used a former Golden Eagle eyrie inland, about 450 metres above sea level, the highest situation amongst the Scottish Sea Eagles. This chick was satellite tagged. Five other first-time breeders, a little older, also proved successful, one of them with a set of twins. No fewer than ten chicks fledged in Mull alone.

One of the 2010 Mull chicks, a male called Kellan, was reared by a ten-year-old female YBS who had first paired with a fifteen-year-old male in 2004. As they had done over several seasons, they catered for their single fledgling for several more weeks until, in September, Kellan was found in a bad way by a local farmer. Dave Sexton managed to catch him and deliver him to the SSPCA vet Romain Pizzi, based on the mainland. The bird was found to have a broken wing, hairline fractures on one leg and a massive bone infection. Not surprisingly Kellan was severely underweight but the vet deemed it worth a try to save him, with 'no guarantees'. Remarkably the vet succeeded so that three months later, on 22 December, Kellan was able to be released back in Mull. On his first flight he was 'greeted' by an adult eagle – his mother, YBS! Dave and his assistant Debbie Thorne provided food over the winter for the young eagle and in April 2011 they found him feeding at a deer carcass that he had found for himself. He looked in good condition and although his radio tag gave up soon afterwards, he was easily recognisable by a slight kink in his right wing. He survived some severe May storms and was seen fit and well in July 2011. Kellan's story is a remarkable one and a tribute to the vet and everyone else who saw him through his serious injuries.

2011 would see 45–52 clutches laid, from which at least 40 young fledged. This figure could have been a bit higher but some nests are proving difficult to monitor accurately each year. 2012 proved even more successful, with 60 young fledged out of some 66 territories established. There were by then (September 2012) 12–14 pairs in Skye and Lochalsh (seven young fledged), three in the Small Isles of Canna (two young) and Rum, 18 in the Outer Hebrides (20 young), five in Wester Ross (three young), seven in Lochaber and mainland Argyll (six young) with at least 16 in Mull and the Argyll Islands (22 young fledged).

To assess the fortunes of the Scottish population so far, four clutches might have been taken by egg collectors in early years while only six of the 25 Sea Eagles recovered dead in Scotland up to 2004 (before Phase III) had been victims of persecution, some of them poisoned. Being so fond of carrion, sea eagles – especially the inexperienced young – are particularly at risk from illegal poisoned baits. Even worse, however, was the fact that two thirds of the poisoned birds were adults. Losing established breeding birds with their prospect of a long and productive life is highly damaging to reintroduction efforts.

An adult Sea Eagle in the Hebrides

On the other hand, survival rates and breeding success have been remarkably good. Analyses published recently by my colleagues (Whitfield et al. 2009) have revealed that survival is now 74 per cent for first-year birds and 97 per cent for adults, comparing favourably with wild populations elsewhere. Our oldest birds are now over 30 years of age. As expected, males breed slightly earlier, at 4.8 years compared with 5.4 for females, a younger age than in other populations, perhaps because Scotland offers so much unoccupied habitat. With slightly more males than females imported or surviving at the outset, within such a small population, breeding was slow to kick off. It is also possible that low densities made it difficult for birds to locate each other. Indeed, some territories were at first occupied by trios (invariably one male and two females) but such cases are now rare or non-existent.

At 34 per cent breeding success, with 0.55 young fledging per breeding attempt and a brood size of 1.48, production was lower between the years 1982–1992. During the period 1993–2000, though, brood size would average 1.61 with more but younger pairs being recruited into the expanding population. Similarly, breeding success improved to 49 per cent with 0.72 young being fledged per breeding attempt. Productivity has increased steadily with 0.76 young per territorial pair in recent years. Perhaps due to the lack of parental care and example, Sea Eagles released during Phases I and II showed slightly lower breeding success than wild-bred pairs and as more of the latter are recruited, production will continue to improve. This now compares favourably with some current European countries, certainly in our donor Norwegian coastal population. The recovering Swedish population, with a higher density, does better.

I am proud to say that Site 17, near my home, remains one of our best breeding pairs to date. A female with a young male had been in the area for a year or so before they laid their first clutch in 1995 but it would be another year before they reared their first chick. We think the female was 13 years old, having been released in Rum with yellow wing tags in 1983. The male, with green wing tags, was only four years old and had been bred in the wild in 1992. At the time, he was the youngest bird known to have reared a youngster in our population. They failed in 1997 but produced their first twins the following season. They went on to produce another 11 sets of twins and only four single chicks, with another failure in 2008 when I think a new female had taken over. In all, up to 2012, this territory has fledged an impressive total of 26 young in 19 years.

There remains the question of competition with Golden Eagles which, in many cases, had moved in to occupy coastal sites occupied by Sea Eagles in the past. The Golden Eagle tends to be a more aggressive bird than the Sea Eagle and has been known to oust the latter from food dumps (Halley and Gjershaug 1998), and even from eyries (Willgohs, Helander and others). I discussed the situation at length, first in *The Return of the Sea Eagle* (1983) and then in a presentation to the Sea Eagle 2000 conference in Sweden (Love 2003). In the first 18 months after the first Sea Eagles were released in Rum I observed 15 attacks on the new arrivals by Golden Eagles. But in the subsequent six years I recorded only seven. In 1975 four pairs of Golden Eagles regularly nested in Rum, some of them on coastal cliffs which were doubtless old Sea Eagle sites. When the first Sea Eagles tried to nest in Rum they chose one site close to a coastal Goldie pair which, having suffered poor breeding success for some years, promptly left. Another Golden Eagle pair moved inland. The 'laid-back' Sea Eagles were obviously capable of holding their own.

Two pairs of Golden Eagles nested on the neighbouring island of Canna at that time, feeding on seabirds and the abundant rabbits. Rum has no lagomorphs so the young Sea Eagles were soon attracted to Canna. One of the Golden Eagle pairs left altogether (temporarily nesting in Rum, with its seabirds and abundant deer carrion, and bringing Rum's total to five pairs for a couple of years). There are now two pairs of Sea Eagles nesting in Canna – although one occasionally uses an alternative site in Rum – with one pair at either end of the

The exceptional breeding success at Site 17

10 km-long island. The remaining pair of Golden Eagles nests in between them, less than half a kilometre from one of the Sea Eagle nests.

Ken Crane and Kate Nellist describe what they considered surprisingly few encounters between the species in the Isle of Skye in their fascinating book *Island Eagles* (1999). On one occasion seven young Golden Eagles (later joined by an adult) were interacting in the air together for an hour, while two immature Sea Eagles:

> 'true to their lethargic nature, perched on the hill…quietly watching the whole event. Eventually they joined the other birds and they all continued circling, soaring and engaging in mock attacks in groups of twos and threes throughout the afternoon.

At another Golden Eagle eyrie, also in Skye, an unpaired adult Sea Eagle was perched nearby, unperturbed by the occupants' efforts to make it move on. Twice the Sea Eagle even landed on the rim of the eyrie, once displacing the female Golden Eagle. A particularly persistent pair of Sea Eagles even seem to have ousted a long- established pair of Goldies, both birds Ken and Kate reckoned to be quite old. Eventually the male Golden Eagle was found dead, partly plucked and eaten, below the nest crag with the Sea Eagles in residence. Post mortem examination indicated the victim had taken a blow to the head. Maybe the Sea Eagle is not as easy-going as we think? Bjørn Helander told me of two cases in Swedish Lapland where Sea Eagles took over nests that had been occupied by golden eagles the year before.

The two species co-exist in Scandinavia and elsewhere to this day, as once they did in Scotland too. Recent research on diet now shows that there is less overlap in diet here than was once thought (Whitfield *et al* 2013). In Rum I had noticed how Sea Eagles hunted on the coast with Goldies foraging more inland (Love 2003). Whitfield *et al.* (2002) and then

Ken Crane and Kate Nellist who have monitored eagles of both species in Skye for many years (J A Love)

Evans *et al.* (2010) compared the two species in the west coast of Scotland in more detail. They found White-tailed Sea Eagles prefer trees if they are available, compared with the cliff-nesting west coast Golden Eagles. Sea Eagles therefore tend to nest in more open wooded habitats and at lower altitudes than Goldies and prefer open water nearby. In addition, Sea Eagles take a wider range of prey which, as might be expected, includes more seabirds and fish than Golden Eagles which favour a lot of rabbits and hares. Sea Eagles might fly several kilometres to seabird colonies, even crossing Golden Eagle territories. They seek out wetland or coastal habitats in which to hunt, while Goldies head for the hills, above the tree line. More and more evidence is accruing to indicate that the two species are partitioning up nest sites and food resources according to their own species preferences. There is plenty of space for both.

So, in summary, to date, including both Scotland and Ireland (each with at least one more year of imports to come), 319 imported young have so far been released (Appendix III). Up to 2012 a total of 467 young Sea Eagles have been fledged in the wild in Scotland (Appendix IV). It is a delight to report that, in April 2012, a pair of Sea Eagles incubated eggs (albeit unsuccessfully) in Co. Clare, the first to breed in Ireland for over a century. The female is only three and the male a year older, which is indeed a remarkable turn of events. With the number of territories increasing each year, together with the number of young fledged in the wild, the future for the Sea Eagle, back home in Britain and Ireland at last, looks bright indeed.

Chapter 12
Living with Sea Eagles

> I have followed with great interest the progress of this project since these magnificent birds were brought over to Rum from Norway. The reintroduction of the White-tailed Eagle to Scotland is a superb achievement and a tribute to many organisations and individuals who had the passion, ability and determination to deliver a real conservation success story.
>
> Sam Galbraith, Scottish Environment Minister (2000)

It must be said that the great majority of the British people welcomed the return of the Sea Eagle, which came to be celebrated not just by the general public but by countless politicians, in local communities and on national television, but also in conservation circles both at home and worldwide. It seems that only a few vociferous opponents were highlighted by the press to focus on a single negative issue – that of lamb killing – because confrontation between conservationists and crofters always makes a good story.

Today it seems to remain a common misconception that lamb killing only became an issue in the late 1970s with the reintroduction of Sea Eagles. Crofters and shepherds seem to have forgotten that, prior to that, they directed their complaints at Golden Eagles. As a result there were some early studies into the impact of Golden Eagles which are still relevant to our arguments here. There are now many more Golden Eagles than when legislation was first introduced, with the Protection of Birds Act in 1954, but Golden Eagles seem to have dropped out of the limelight. Now that Sea Eagles have been brought back into the equation, there is a new figure – the 'conservationist' – to whom crofters and critics can direct complaints. Grievances are usually pursued with such vehemence and hyped up by the media to such a degree that all reasonable perspective is lost. 'Crofters are now a rare species too and who is protecting them?' is a common mantra. If anybody had bothered to consult the crofters, they claim, there would have been unanimous disapproval. All too soon it becomes difficult to hold a reasonable debate and all objectivity is lost.

Clearly, in advocating the reintroduction of the White-tailed Sea Eagle to our shores, claims of lamb-killing in the past must be critically assessed. One's stance will differ depending on whether one is a sheep farmer or a conservationist, but as an ornithologist I shall attempt to approach the controversy with as much scientific objectivity as I can. As I see it, four considerations have to be borne in mind:

The relative importance of lambs in the diet of the Sea Eagle

The number of lambs which were taken alive by the eagles or as carrion

The proportion of those lambs killed that would have died anyway

The significance of this eagle-induced mortality relative to total lamb deaths in the whole flock

With the Sea Eagle exterminated from Britain we had to take a rather indirect approach to the first issue, either by studying the writings of people who had direct experience of Sea Eagles and lambs in the 19th century, by considering the situation abroad where Sea Eagles are still to be found or by extrapolating from the habits of a similar species – in our case, the Golden Eagle – still common in Scotland.

Around 1960 Dr Jim Lockie (Lockie 1964; Lockie and Stephen 1959) found lamb-killing by Golden Eagles to be a local problem only, prevalent in those areas of northwest Scotland where the eagles' preferred prey – grouse and hares – was scarce. In a survey of the old ornithological, zoological and sporting literature I have gleaned a total of 141 prey items which had been recorded at eyries of the Sea Eagle prior to its demise in Britain. Only 12 (9 per cent) were lambs. Seven of them were found at one nest in the Lake District, serving to indicate how the problem is often of a local nature. Indeed there are several Sea Eagle pairs – St. Kilda, the Shiants, South Uist and Fitful Head in Shetland, for example – which merited specific mention in the literature because they never touched lambs. At the Shiants, Martin Martin (1716) was told by the natives:

> ...that these [Sea] Eagles would never suffer any of their kind to live there but themselves, and that they drove away their young ones as soon as they were able to fly. And they told me likewise, that these Eagles are so careful of the place of their abode, that they never yet killed any Sheep or Lamb on the Island...

When Lord Lilford was travelling through Greece in 1860, he made to raid one Sea Eagle eyrie but 'the shepherds begged us not to kill them as they bred year after year and kept away other birds of prey which were destructive to their lambs'. With a knowing smile Lilford was disinclined to believe them but, to his credit 'scrupulously attended to their request'.

Back on home territory, there are of course frequent but unsubstantiated statements about how destructive Sea Eagles are to lambs. There are remarkably few accounts, though, of the birds feeding on lambs or sheep, let alone actually killing them. I uncovered a further 46 observations of Sea Eagles feeding at prey away from the nest, mostly in the winter months and sheep and lambs feature in only six of them. Four involved eagles feeding at adult sheep, which must certainly have been encountered as carrion. One lamb is recorded as being 'lifted' by a Sea Eagle, but it was immediately dropped again unharmed. Another instance is detailed where 10–12 lambs were picked up one after the other but each dropped at once, again all unharmed. Eagles on one small island in Shetland were accused of killing seven lambs 'in a short time'. Two other accounts were total fiction, a good story perhaps but almost certainly not true.

I am not claiming that Sea Eagles cannot kill and have not killed lambs. Some references to eagles killing lambs are undoubtedly based upon fact and, to be fair, we must remember that the very shepherds who were the most likely to witness such depredations on their flock were the least likely to record their experiences on paper.

But it can be all too easy to overstate the problem. Studies in Norway and elsewhere in recent times would certainly bear this out. Willgohs (1961) gathered 36 instances of lambs or goat kids being lifted by eagles but only 12 could be said to have been killed by the eagles. In a further five instances it was positively known that the animals had been taken dead. Many of the reports emanated from the southern counties of Norway, with relatively few further north where eagles were much more common and where it was frequently intimated to Willgohs in the course of his researches that the eagles were harmless. In a list of over 1400 prey items noted from Norwegian eyries, only 60 (4 per cent)were sheep and lambs, whether live or dead, so these are of very minor importance, numerically, in the diet of the Sea Eagles. Harald Misund agrees that the preying of Sea Eagles on lambs is just not an issue in northern Norway. Indeed, I have seen small flocks of sheep and lambs grazing quite happily through open scrub woodland while eighteen immature and subadult Sea Eagles roosted in the branches directly above.

Beinn Eighe and the Torridon hills where Jim Lockie did his Golden Eagle/ lamb studies in the 1960s

Although to a crofter a lamb is a lamb, it would be wrong to accord all prey items equal value to an eagle. In terms of bulk, lambs assume greater importance since they are so much larger than most other prey items. In Wester Ross, where Golden Eagles were taking an unusual number of lambs, Jim Lockie (1964) calculated that they comprised some 30 per cent in weight of the eagles' intake. He found remains of 22 lambs in one eyrie, over five consecutive seasons, and on ten of the carcasses he was able to carry out a useful post mortem; seven had been alive when lifted by the eagles, as evidenced by bruising and bleeding under the skin where it had been pierced by the talons. This can be viewed in clearer perspective when one considers that there were 1,000 ewes breeding in the territory of that eagle pair. In 1973 the Hon. Doug Weir (1973) estimated that another pair of 'rogue' Golden Eagles in the Highlands had taken some 20 lambs during the period that their eaglets were in the nest; this made up some 75–85 per cent by weight of the prey found at their eyrie but even in this instance, it represented only some 2 per cent of lambs born within the hunting range of that eagle pair. Only two of the four carcasses examined could be claimed to have been killed by the eagles.

It is also important, of course, to be absolutely clear just how many of the lambs lifted by eagles were in fact healthy and would otherwise have survived. In his definitive monograph on Golden Eagles the late Jeff Watson (1979, 2010) calculated that, at average sheep densities in the Highlands, a typical Golden Eagle range of 40 sq. km might hold about 1,000 ewes, 5 per cent of which would be barren. Despite occasional twins, the rest would produce only some 600–700 lambs. Of these, at current levels, 30 per cent would be expected to die (either stillborn or dying of starvation or disease) or be taken by other predators such as crows or foxes. Golden Eagles alone might kill no more than 2.4 per cent of the total lamb crop, up to 23 lambs per year. On the other hand, most predators readily scavenge dead lambs and there would seem to be plenty available.

In 1977 David Houston also presented data on lamb mortality in Argyllshire, where some 15–20 per cent of lambs died, mostly within their first week of life. He examined the carcasses of 254 of them to determine the causes of death; 27 per cent had been stillborn, 5 per cent had died almost immediately after birth and before they had walked, 9 per cent were diseased and another 5 per cent were killed by accident, having been sat upon by the ewe, strangled in fences, hit by cars or killed by dogs. Among the remainder for which no cause of death was apparent, nearly all of them (45 per cent of the entire sample of 254) were found to have exhausted fat levels, meaning that they had effectively starved – 75 per cent of them had never even sucked. A few of the others had 'reduced' fat levels and only 7 per cent seemed healthy. This sad reflection on sheep husbandry in some areas of the Highlands would appear to be by no means limited to recent times. As long ago as 1871 Robert Gray accepted that:

> . . . it cannot be denied that the Sea Eagle oftener feasts upon carrion than upon living animals, and that in most of cases where lambs are actually lifted the offence is to a great extent mitigated by the fact of the severe spring weather having previously crippled these poor creatures beyond hope of recovery. . .

It is well known that the age and health of a ewe can also affect her lamb. She may be unable to produce enough milk to feed it, or the lamb may be small or weak and therefore more susceptible to exposure. Better-fed ewes, on the other hand, are more aggressive in defending their offspring from predators. Regular feeding of ewes and the practice of lambing on richer pastures were found to reduce mortality in blackface lambs from 14 to 5 per cent. 'Flushed' ewes, given extra feeding in the weeks prior to their being put to the ram, tended to have a higher incidence of twinning. Although twin lambs may be smaller and more susceptible to exposure, such enhanced breeding output means than one of the twins can easily be cross-fostered on to another ewe which might have lost her lamb. Such improvements in sheep husbandry may not, however, diminish the frequency of stillbirths. Indeed, difficult births may be more likely to occur if the lamb is a large one so that, in addition to the possibility of the lamb dying, the ewe herself may be put at risk.

Jane Hall, a research student, responded to complaints of Golden Eagles killing lambs in North Uist in 1976; she found few remains in the six eyries she visited so could not determine if the claims that eagles were actually killing lambs were true. Natural prey seemed scarce and the eagles' breeding success that year was poor, the lambing percentage locally being only 65 per cent with evidence in some places of overstocking. Back in the 1980s Alan Leitch

The naturalist William Macgillivray (1796–1852), brought up in Harris and later Professor of Natural History at Aberdeen University

(1986), who studied Golden Eagle predation in Glenelg, had concluded that a relatively small investment in improved husbandry would improve ewe condition and thereby increase lamb production beyond even the highest levels of predation by eagles. Such enhanced techniques (and to this must be added careful maintenance and improvement of hill pasture, together with rigorous culling by the shepherd of animals showing poor performance) will all help increase lamb survival. It seems to me that benefits accruing from such efforts far outweigh any brought about by time-consuming – and highly illegal – predator control.

In the United States it has been demonstrated (Wagner 1972) that extensive poisoning of coyotes – a species considered by many American farmers to be a significant menace to their sheep – made no difference whatsoever to the annual losses of stock. This is in accordance with an idea proposed by the pioneering ecologist Aldo Leopold as long ago as 1946, by which:

> ... a great deal of predation is without truly depressive influence, in the sense that victims of one agency simply miss becoming victims of another...

With losses due to other causes being so high in some parts of the Highlands, the abundance of carcasses would largely preclude the need for many lambs to be killed by predators. Lockie noted that lambs featured more in the diet of some Golden Eagles in years of poor lamb survival. In good lambing years, on the other hand, the incidence of lamb remains found at eyries was halved, from 46 per cent to only 23 per cent.

It might be significant that the naturalist William Macgillivray, brought up in Harris, had more to say about lambs when he was discussing the Golden Eagle in his book on raptors (1836). But on the same theme, the late Dr J.W.Campbell had once commented:

> One is left with the impression that the threat normally to sheep under average conditions, was not greater from the Sea Eagle than it is today from the Golden.

I would even claim that it is much less, because of the Sea Eagle's decided preference for carrion. Dixon (1900), writing from the Gairloch area, astutely remarked that of the two species the White-tailed Sea Eagle:

> ... is a regular scavenger of the shore...Healthy vigorous birds or animals are seldom attacked by this eagle; it confines its attentions to the weakly and the wounded creatures that cannot move fast or offer any serious resistance.

On the question of livestock, William Macgillivray (1836) was more specific:

> ... at seasons of mortality among sheep, as in the end of autumn, when braxy commits its ravages, or in the end of spring, when severe weather often causes the death of young lambs, they {Sea Eagles] are not commonly seen hovering about.

Such was the situation in the early days of the reintroduction project and when *The Return of the Sea Eagle* was published. It is curious to reflect how, prior to the Sea Eagle's return, animosity was entirely directed at the Golden Eagle. This, though, was protected by law so there was little hope of getting officialdom to react.

Only once the agricultural community, egged on by the media hype, 'realised' that Sea Eagles were being reintroduced, first by a government agency (NCC/SNH) and subsequently by a conservation body (the RSPB), did they see an easier target. The Sea Eagle is also protected but complaints against them generated further critical studies. At the behest of, and indeed with the co-operation of, worried crofters and farmers, more detailed studies were made by independent researchers first in Mull (Madders and Marquiss 1999; Marquiss and Madders 2000; Marquiss, Madders and Carss 2001; Marquiss, Madders and Carss 2002; Marquiss, Madders and Carss 2003; Madders and Marquiss 2003; Marquiss, Madders, Marquiss and Carss 2003a; Watson, Leitch and Broad 1992), later at other Sea Eagle territories throughout the west of Scotland and, latterly, in the Gairloch area (Simms *et al.* 2009). The result is that more is now known about the diet of reintroduced Sea Eagles than of any other raptor population in Britain. To anyone who wants to know about Sea Eagles and lambs, I would recommend any of the scientific papers and reports that have appeared recently.

From my analysis in Chapter 4 it is blatantly obvious that, with such a diverse range of hunting and foraging techniques at its command, the Sea Eagle possesses a very catholic diet, taking a wide range of bird species. It takes a wider range of prey, in fact, than does the Golden Eagle. Although there is considerable overlap where the two species are neighbours, the Golden Eagle takes many more grouse and lagomorphs while the Sea Eagle kills more seabirds (especially fulmars) waders and – commonly hunting along the coast – will, of course, take fish, something denied to golden eagles. Both species take carrion.

I am not aware of any recorded instance of Sea or Golden Eagles killing full grown sheep; it is hardly possible that they could do so. However eagles of both species readily scavenge upon dead sheep so it is not surprising that wool is often found in eyries and in pellets. We have never disputed, despite claims to the contrary by critics and the media, that lamb remains have been found in Sea Eagle eyries in Scotland. Many Sea Eagle eyries in the Highlands and Islands will reveal evidence of lambs, some more than others. But no one could condemn Sea Eagles for scavenging dead lambs, and indeed such carrion-eating birds fulfil a useful sanitary function if carcasses are simply left out there for the taking.

Adult Sea Eagle

It is the question of whether the eagles found these lambs already dead or had actually killed them that requires careful scrutiny. If a lamb has had its eyes pecked out, then crows or Ravens got there first. A Sea Eagle would never go for the eyes so it must have found such a lamb already dead. A new-born lamb emerges with membranes over its hooves which then quickly rub off when it first walks. If these membranes are present then the lamb was stillborn when the eagle found it.

Certainly evidence of killing can be quite hard to come by. One of the only sure ways to identify an eagle kill is to find puncture marks from talons under the skin, usually on the lamb's back or shoulders and associated with bruising and/or bleeding. Not all lamb remains will present such evidence. Direct observation of attacks or kills is another indication, of course, but this is more rarely witnessed. Shepherds are perhaps best positioned to see interactions between eagles and lambs and it was from one corner of Mull that the issue first came to the fore.

Sea Eagles lifted most lambs in May, when most are being born, with fewer to be found in eyries in June and almost none in July and August when the young eagles are fully grown and fledging, and presumably when the lambs are too big. Blackface lambs weigh about 3 kg at birth. By August the lamb will be around 25 kg, probably well beyond the weight that a sea eagle can normally carry; the carcass would need to be dismembered first.

In their classic studies Mick Marquiss and Mike Madders assessed those carcasses that could be identified as either having been killed or scavenged. One nest in Mull presented the worst case scenario but, it has to be stressed, only in one year of the study. Of the 45 lamb remains taken by this pair in 1999, 21 were intact enough to show the cause of death and revealed that 13 (or 62 per cent) had been taken alive. This suggested that they might have killed a total of 30 all season. Rabbits had been scarce that year due to an earlier myxomatosis outbreak. In the year 2000 this pair moved to another nest near the coast where fewer lambs were available but rabbit numbers had largely recovered. Thus of 19 lambs examined in the nest only four (21 per cent) showed signs of predation.

In 1999 a second pair had scavenged all but one of the 11 lambs examined. No data was available the following year since their breeding attempt failed. The third Mull pair investigated that year took hardly any lambs despite having to fly over lambing areas to catch seabirds on the coast 15 km distant. In 2001, at a nest in Skye where the highest total of lambs was found, all had been scavenged and none killed.

Finally, account has to be taken of the vulnerability of lambs to predation, their state of health and whether the killed lambs might otherwise have survived. It is recognised that ewes can find it difficult to protect twins and some shepherds prefer single lambs as being heavier and healthier than twins. A ewe's ability to protect her lamb or lambs might be impaired by her own poor health. Furthermore, poor nutrition, disease or tick infestations might impair the ability of a lamb to follow her mother adequately. Certainly some lamb carcasses recovered from eagle nests during the Mull study carried heavy tick infestations. Furthermore the lambs consumed by Sea Eagles, whether diagnosed as killed or scavenged, tended to be small for their age, resembling those that can be found lying on the hill having died from natural causes.

It was not easy to gauge the health status of lambs lifted by eagles but there is a suggestion, in common with other studies of predators and their prey, that at least a proportion of those lambs killed by Sea Eagles were in poorer condition and might not have survived anyway. Without lengthy and expensive chemical analysis of fat deposits or bone marrow, Mick

Marquiss sought an easier alternative method of assessing the health status of the lamb remains he found. While the length of leg bones gave an indication of age, it was local crofters who pointed out to him how the length of the fleece on its back can reflect the condition of a lamb. Armed with this useful observation Mick was able to show that a significant proportion of lambs found in Sea Eagle nests were of poor quality and, even if the eagle had not killed them, might not have survived anyway.

The Mull studies went on to reiterate how actual eagle predation can be minimal when compared with other factors resulting in lamb deaths. On relatively poor hill ground in Mull the number of lambs produced per 100 breeding ewes (a lambing percentage of 60–70 per cent) can be half that experienced on intensively-farmed units on lowland. It was estimated that Mull supported over 20,000 breeding ewes in total. When the study was widened to include data from the other seven nests in Mull at that time, the annual consumption of lambs, mostly in May and June, was in the order of 200 but, crucially, most of them would have been scavenged rather than killed. Compared with other eagle/livestock studies in Scotland and elsewhere, the Mull study confirmed that the numbers of lambs killed varied between pairs and from year to year. Furthermore, the losses attributable to eagles were small compared with other sources of mortality, with minimal impact to the sheep economy of the island as a whole.

A more recent study was undertaken to investigate alleged lamb losses in the Gairloch area (Simms *et al.* 2009), but none succumbed to eagles in 2010, the year of the study. The crofters were all too quick to counter that it was probably the activities of the researchers in the lambing park that had deterred the eagles. In the summer of 2011 I was invited by the local community to give a lecture on Sea Eagles. It was very well attended but not one local crofter came along to hear the true story. Only one man who had come all the way from Lochinver initiated a lively but brief debate at the end. He took on board what I said and we parted good friends.

Poolewe, Wester Ross, where there have recently been complaints of Sea Eagles taking lambs (Noble Caledonia)

. only specific Sea Eagle pairs were inclined to kill lambs so any impact was highly :ven within the island. This, of course, could prove a significant loss to the individual cro... small farmer being targetted. The idea of compensating for lambs lost has always proved costly, futile and open to abuse. I am told that farmers used to be compensated for cattle that were struck by trains until it was realised that in some places any cow that keeled over, for whatever reason, was dragged to the nearest railway line so that compensation could be claimed. A Norwegian I met recently explained how Lapps were being paid compensation for reindeer calves alleged to have been killed by wolverines. Eventually the authorities did their sums and realised that the entire population of wolverines in Norway – and there are not that many of them – could not possibly kill and eat that many reindeer.

An alternative approach might be to deter eagles from entering lambing parks but how this might be achieved requires more investigation. Certainly bringing lambing closer to buildings can be enough to deter eagles, while providing some sort of shelter might offer refuge from attack. In the USA the removal of the offending Golden Eagles altogether has been shown to be expensive and ineffective, as others quickly move in. However, there are factors that should be within the immediate capability of the crofter to address and that could help reduce losses.

Studies indicated that lamb mortality in Scottish hill sheep is increased by poor nutrition. In the ewe this leads to difficult birthing, poor milk yield and poor parental care. This not only results in lamb deaths at birth or from starvation but also predisposes them both to predation and probably to disease through reduced immune responses. Thus a crofter exposed to undue Sea Eagle predation on his lambs, rather than being compensated for losses, could be subsidised to improve the nutrition levels in his flock and to reduce the levels of tick infestation. Furthermore, moving his lambing ewes away from the vicinity of eagle nests or closer to human contact might also show benefits.

It was this very approach that led to management schemes being put in place in Mull and elsewhere. These seem acceptable to sheep farmers, if only as recognition that a problem can exist. And of course, additional sources of income always prove attractive, especially when returns from sheep sales are so low. Undoubtedly, such measures also help to maintain viable rural communities and their culture in the Highlands and Islands and legitimise the concept that agriculture in less favoured areas can be as much about subsidies for conservation as for food production.

Finally, it may come as a bit of a surprise to hear that sheep farming is carried out in southwest Greenland where some 50 farms run about 20,000 sheep, alongside about 150–170 pairs of White-tailed Sea Eagles. The marginal climate and conditions resulted in quite a high mortality and the eagles were being blamed for losses. Studies have shown, however, just how rarely lambs feature in the birds' diet. Persecution still occurs but a determined conservation effort has overcome the problem to a large extent, while showing just how marginal the area is for sheep rearing. There was even talk of raising the Sea Eagle to the status of Greenland's national bird.

It has to be said that the level of attack upon the project over the years – and especially upon the individuals involved in it – has been wearisome indeed. It rose to the surface again as recently as June 2012, bringing up the same old, usually false, even ridiculous claims, although this time one critic even invoked the Sea Eagle in the decline of the commercial fishing industry in the North Atlantic. I wonder if he has heard of over-fishing or marine pollution?

I do not wish to give our critics or their absurd media coverage any further airing. I will simply quote one exchange of letters in 2005 that involved me personally. For a time

SNH managed to publish an occasional supplement called 'Teachd an Tir' inside the *West Highland Free Press* (which was not always a fan of SNH) explaining their work and providing background to local conservation initiatives. In the second issue (Spring 2005) my article about the Morar eagle poisonings, 'A Sad Saga of Special Sea Eagles', was given front page coverage. I have already presented the whole sad story but in the article I celebrated the very first successful fledging in Scotland of 'the first Gaelic-speaking sea eagle for 70 years'.

This immediately drew a predictable response from a vociferous critic of conservation who, teaming up with another vitriolic author, promotes an organisation called People Too. (A crofter I know said it should be called Two People and I have no idea if their ill-informed crusade lasted very long). The lady in question (I will at least accord her the courtesy of not mentioning her name) despatched a letter to local newspapers. Only the *Stornoway Gazette* (who had no real idea what the correspondent was on about (since the article had not appeared in their paper in the first place), printed the letter. Referring to my 'Gaelic-speaking Sea Eagle' she wrote:

> There is a statement of profound significance from a scientific organisation. Here also is a truly remarkable linguistic state of affairs, given that the eaglet's parents probably spoke Norwegian and that neither baby nor the adults had any obvious direct contact with Gaelic speakers. The assumption must be that an ability to speak Gaelic can be acquired by eating crofters' lambs...

Groan. I could have forgiven the woman lacking any sense of humour but she would not let the issue go at that. She next wrote to the Chairman of SNH – my boss – demanding an apology and an undertaking that I would never use the statement again. She will be appalled to discover that I continue to use the phrase in all my lectures and have done so for years and that it never ceases to raise a laugh from the audience. And if at last she seeks to inform her crusade by reading this book, she may notice I have used the frivolous phrase several times in these pages.

So in summary, here in Scotland as elsewhere, we can see how accusations of lamb-killing have been constantly misrepresented and grossly overstated. Sea Eagles have killed lambs but, if anything, lamb killing is a very local issue. Critics of the Sea Eagle reintroduction not only forget that most lamb remains found in Sea Eagle nests have been taken there as carrion, but that although it may impact individuals in some years, it will never threaten crofting as a livelihood in the Highlands and Islands. Indeed, conservation management inducements are helping to support crofting and supplement crofters' incomes. Admissions to the Sea Eagle hide in Mull even provided grants to the island's agricultural show.

A year or two back Martin Carty invited me to Mallaig to visit a couple of local primary schools and I gave a public lecture in the village that evening, only too delighted to see such a turnout of friends, some of whom had greatly assisted me with the Rum Sea Eagles.

It is gratifying to see such support and to reflect upon the fact that hosts of people throughout the country have been very excited by and hugely supportive of the bird's return. It is now generating its own – not insignificant – tourist appeal, as evidenced by how commonly Sea Eagles are displayed on prints, postcards, mugs and other souvenirs in craftshops all over the West Highlands. Coach tours, wildlife safaris, hotels and guesthouses all proudly advertise their local Sea Eagles and it attracts regular custom – 'bums in beds', as one operator put it.

I particularly enjoyed an engaging programme on the Gaelic TV channel, BBC Alba, featuring the return of the Sea Eagle to Mull. Instead of focussing once again on the lamb controversy, it told of the successful return of the Sea Eagle alongside the return of the ancient language to an island that had fast been losing its native tongue. We were shown how the Gaelic-medium children in Mull knew all about their local Sea Eagles, with Dave Sexton in Mull, Alison Maclennan in Skye, Kenny Nelson in Poolewe, Claire Smith and Rhian Evans with many others doing stirling work with schools promoting wildlife conservation and eagles.

This level of interest clearly demonstrates the support for Sea Eagles amongst the Gaelic community, of all ages, in the West of Scotland. Alan Reid's song about the project on his CD *The Sunlit Eye* has a Gaelic chorus sung by popular Skye artists like Arthur Cormack of the Aros Centre and Christine Primrose:

> ...For sixty long years he was banished and gone.
> They plundered his nest, they hunted him down;
> But now from the land of the fjords he's returned,
> And Iolair na mara is home in the islands again...

The well-known performer Allan Macdonald (one of the three world-renowned piper brothers from Glenuig) has also composed a Gaelic song about the Sea Eagle and many more examples will doubtless appear in the future. There is no clearer demonstration of the Sea Eagle's popularity in the Highlands and Islands, and indeed throughout the length and breadth of the country, than the public viewing attractions.

CD The Sunlit Eye *by Alan Reid*

Back in the 1960s, a significant turning point in showing off and promoting raptors was the invitation to the general public to view the first osprey pair nesting at Loch Garten in Speyside, only a year or two after the first nesting attempts. It was a bold move by George Waterston and its success – even today – cannot be doubted. I spent several weeks there as a volunteer warden at that time and, even back then, was astonished by the amount of interest shown in the facility. It was here I cut my teeth in raptor conservation and first met George Waterston, Roy Dennis, Doug Weir and many others.

The ospreys really put Boat of Garten on the map, bringing huge benefits to the local tourist industry whilst providing a lucrative membership arena for the RSPB. Particular nests of other Ospreys, Red Kites, Hen Harriers, Peregrines and Golden Eagles have since been opened to the public and wildlife watching has become a hugely popular pursuit. The number of sightings reported back to the project over the years more than amply demonstrates just how much enjoyment is being derived from the birds' presence by bird-watchers and the public.

In the late 1970s, as soon as reports of eagles taking lambs began to be made in Mull, I felt that we needed to take a leaf out of George Waterston's book and open a hide to the public.

Dave Sexton and the Mull Eagle Watch (Debbie Thorne/RSPB)

Already the first nests were quite well known and wildlife tourist operators were showing off the birds to their clients. They were respectful of the nests, while such public interest on a small island like Mull meant that egg collectors were usually pretty obvious to everyone. That is not to say that clutches were not stolen occasionally but I feel it did not become quite the problem that we first anticipated. I think eggers were slow to realise that the reintroduced Sea Eagles were now a 'real' target species. Very quickly the folk on Mull began co-operating with the police to mount an Eagle Watch during the vulnerable period of laying and incubation.

It was not until 2000 that the first tentative steps to the public viewing of a Sea Eagle nest were undertaken (Maclennan and Evans 2003). These emphasied the concerns for the birds within the local community, the deep interest in catching sight of them in the wider community and the attraction the Sea Eagles had become to visitors from further afield. By now there were several pairs nesting in Mull but the one that offered the best opportunity for a hide was already pretty public, nesting in a sitka spruce plantation not too far from a forestry track that was popular with anglers, cyclists and walkers. Forest Enterprise (FE), who owned the plantation, proved eager to help and the first hide was a converted caravan equipped with a viewing window and interpretative material. There was a convenient car park nearby but visitors (a maximum of 20 at a time) had to book ahead. The hide operated from late May, after the two chicks had hatched, until early August, after they had fledged. FE, the RSPB, SNH and the Isle of Mull Community Trust supervised the operation and a teacher and naturalist, Joyce Henderson, looked after the visitors, who totalled 1,200 in that first season.

The Mull viewing hide in its early days (Dave Sexton)

Although the pair of Sea Eagles had only been nesting together for four years, and not always in the same tree, they usually succeeded in rearing chicks, clearly demonstrating that they had not been unduly disturbed. Although the parents were often a bit slow to reveal which site they were going to use, the hide soon developed into a more substantial structure with CCTV cameras. It became popular with local schoolchildren, wildlife tours, general visitors, TV personalities, even dignitaries and politicians. Via the Sea Eagle project team annual reports and blogs, Dave Sexton and Debbie Thorne have provided updates. In 2005, despite the introduction of a small charge, the hide admitted over 3,500 people. Half of the proceeds, totalling some £10,000–£12,000 annually and rising, is given to numerous local causes and charities, including the local agricultural show. Mull Eagle Watch hide received a five-star (exceptional) award from Visit Scotland, one of only two wildlife attractions in Scotland to win this accolade. Mull Eagle Watch and the Hide is run by 5 different organisations – Forestry Commission Scotland, Scottish Natural Heritage, Strathclyde Police, Mull & Iona Community Trust, and RSPB Scotland along with volunteers both visiting and from the local community. The award confirms how well they all work together, each adding something different and with the end result – eagles nesting successfully and a great wildlife experience.

The screening of the popular BBC TV series *Springwatch* in 2005 reached out to some 3.4 million viewers, and introduced the parent birds Frisa and her mate Skye, along with their twins Itchy and Scratchy (named by a local primary school), to a massive, nationwide audience. *Springwatch* has since returned, with the eagles being regularly acclaimed in other programmes such as *Natural World*, *The One Show* and *Landward*. The open season at the Mull hide was extended so visitor numbers in 2006 exceeded 5,500. In 2008 the Scottish First Minister opened a new, improved hide, provided by the Forestry Commission, with more and more politicians wanting to be seen with Sea Eagles on national television.

An adult Sea Eagle on the cliffs near Portree in Skye (Bob McMillan)

The birds have long been a similar attraction in Skye. Sea Eagle eyries here tend to be on more remote and unsuitable cliffs, so a different approach was adopted, again in the year 2000 (Maclennan and Evans 2003). In partnership with Aros (Isle of Skye) Ltd, RSPB Scotland, SNH, Highland Council and the local Enterprise Company, CCTV cameras were installed close to the cliff nest of a long-established pair. Live pictures could then be transmitted back to a manned interpretive facility in the Aros Visitor Centre near Portree. The trial, supported by Forest Enterprise, the landowner and the local crofting township, was limited to the month prior to the chicks' fledging, but that first year the facility attracted 1,200 visitors. It would peak at 9,000 only four years later (Alison Maclennan, pers. comm. and project team annual

Shieldaig

reports). The parent birds, both released on Rum in 1982, had produced a total of 21 chicks over 13 breeding seasons. In 2006 they hatched three chicks, only the second instance of triplets in the project at that time, although sadly the youngest died three weeks later during a spell of bad weather. They have, however, proved somewhat erratic in their annual choice of nest site, while unseasonal bad weather can introduce technical problems and occasional breeding failures, so the Aros facility has become a more costly and time-consuming venture. If there were no eagles to be seen, though, cameras were turned on a local Heron nest, with a webcam on a Sea Eagle eyrie in Estonia or Norway. Everybody still seemed happy and Aros continues to be popular, pulling in up to 5,500 visitors a year. This should rise after the exhibit is enlarged and improved.

The beautiful little fishing village of Shieldaig in Wester Ross now has its own resident Sea Eagles which are often to be seen roosting atop the trees on an island just offshore. Wildlife boat trips are a popular local attraction but to sit outside the local pub, a malt whisky in hand, looking at Sea Eagles is also an experience to be relished. The community recently installed a CCTV camera at the usual nest in trees on the island but the eagles moved in 2012 so, frustratingly, the cameras were left looking at an empty nest; hopes are high for next season. (Emma Livingston, Torridon Sea Tours, pers. comm.)

Skye has also provided an opportunity for viewing Sea Eagles in the wild, with several tripper boats operating out of Portree. These are encouraged to adhere to a code of good conduct, so as not to disturb the eagles, and the birds do benefit from fish thrown out for them. The operators soon learnt to insert wine corks to make the fish float, to be discarded while the birds fed, until it was realised just how many were accumulating at the feeding perches and even in the nest. The practice is now discouraged by the code. In Lochmaddy,

North Uist the late Dr John Macleod delighted in taking friends (myself included) out in his boat to show off 'his own' pair Jessie and Hamish. Being a medical man, he had perfected a technique of injecting air into the fish's body cavity using a huge syringe, so that the fish floated. Mull also offers wildlife-watching boat trips.

But such boat-based encounters are not just for the enjoyment of tourists. School children benefit from the experience and ever since the earliest days of the Rum reintroduction, I have been told by fishermen how they too delighted, and still delight, in seeing Sea Eagles come down to discards or even whole fish, causing panic among the ever-attendant seagulls. Anglers thrill at encounters with the Sea Eagles and, along with fish farmers, have never considered them a threat. At least one shellfish farmer in Mull rejoiced at the presence of Sea Eagles since they kept Eiders away from his mussel ropes, and even the crusty TV presenter Jeremy Paxman has been known to wax lyrical about a Sea Eagle he encountered while he was on an angling holiday in Mull.

A recent study by the Progressive Partnership, commissioned and published by the RSPB in Mull (2011) has revealed how an astonishing £5 million is attracted into the island annually due to the reintroduced Sea Eagles. Indeed the tourism generated by the birds also supported, directly or indirectly, some 110 full time jobs. These calculations were based on a survey of more than 1,200 people who visited Mull in 2010. Almost a quarter admitted that the eagles were an important factor in their choice of Mull as a holiday destination.

A similar claim might be made for Skye and other Hebridean or Highland destinations. On a national scale the Scottish Government has estimated that wildlife tourism is now worth over £276 million to the country's economy, supporting 2,763 jobs in the sector. Sea Eagles have certainly contributed to this growth industry.

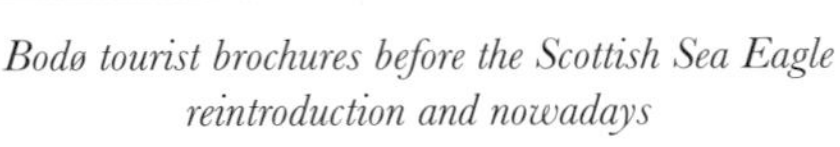

Bodø tourist brochures before the Scottish Sea Eagle reintroduction and nowadays

The Bodø Sea Eagle Club

Harald Misund with sculptures in Bodø town centre

And the benefits of the Scottish Sea Eagle Reintroduction Project are not confined to our own country. It could be argued that it was the donation of four young chicks to Fair Isle in 1968 that gave added impetus to local conservation demands to protect Sea Eagles in Norway, one of the last countries in Europe to do so. Since I first visited Bodø in 1975 I have also noticed how promotion of the town has largely shifted away from its location in the land of the midnight sun and towards its local Sea Eagles.

Postcards and T-shirts for sale in Bodø shops

Thanks mainly due to the efforts of Harald Misund, the NAF base and local landowners, Bodø is justly proud of the part it has played in the Scottish reintroduction. Its logo tends to be a stylised Sea Eagle rather than an uninspiring red sun disc. Artistic images of Sea Eagles are portrayed on T-shirts and mugs, with dramatic photographs of the bird on postcards, prints, calendars and the like. Local businesses have sponsored the Havørnclubben where, for a small registration fee, visitors receive

Sea Eagle sunset

Baggage reclaim at Bodø Airport

avisa Nordland

i dag 108 SIDER

Kjære leser!

LØRDAG

Laget utstilling fra sykesengen

I tre måneder har Nina Thue Nordvik (16) ligget på Nordlandssykehuset. Noe av tiden gikk med til fingermaling, og nå har hun utstilling på rommet sitt.

kultur side 24 og 25

Havørna slår ut Glimt

Over halvparten av Bodøs befolkning mener havørna er aller best egnet til å gjøre byen bedre kjent utad. Det er klart mer enn både Bodø/Glimt, midnattssola og flybyen. Det viser en spørreundersøkelse i regi av Avisa Nordland.

endelig helg side 6-9

Ingrid - ei skikkelig bølle

Ingrid Lamark leder bøllekurs og sitter i landsstyret til Rød Ungdom

endelig helg side 17

Kunne vært reddet

Nordlandssykehuset tar massiv selvkritikk og konkluderer i en internrapport med at Svein Solbakken kunne vært reddet med riktig hjelp da han skadet seg i Skrova 5. juli i år.

nyheter side 9

Foto: Bjørn Erik Olsen

Sommeren er over men vi har varme tilbud hele året!

6085,-

8845,-

Opplev vakre Gran Canaria og lei en bil i 3 dager fra 490,-. Golf på Gran Canaria med "Greenfeepakke 2", fra 2650,-

Star Tour

Se www.startour.no, ring 08100 eller ditt reisebyrå

I DAG ÅPENT TIL KL. 20.00

Coop Mega Rønvik
Coop Mega Mørkved

nyheter side 8

Bondens marked

Lokalproduserte matvarer på Bodø Torg I dag fra 10.00 - 15.00

Headline in local Bodø newspaper where Sea Eagles top the popularity poll, beating the local football team

a colour print by local artist and scientist Jon Fjelsa, a silver pin and a certificate. Even when you step off the plane at the airport and go to pick up luggage from the conveyor belt you are welcomed to 'the realm of the Sea Eagle' by large displays on the wall. There are now several impressive sculptures and statues scattered round the town centre, while for a time a local brewery produced a Sea Eagle beer.

What I think said it all, though, was a poll in the local newspaper, asking its readers what they thought was best about their town. A local football team called Bodø Glimt does reasonably well in the League but the result of the poll, headlined on the front page, was 'Havørna slår ut Glimt!', meaning essentially that 'Sea Eagles are better than our football team!'

So the spin-off benefits of Sea Eagles in Scotland are many, various, local, national and, on an even wider-reaching level, economic as well as conservationist. Who can now argue that the reintroduction was not a good idea?

Over the decades, many people have assisted the project: bemused aircrew on both sides of the North Sea, helpful fishermen anchoring in Rum overnight to offer fish for the eagles, the general public, bird-watchers, officials in numerous agencies both at home and abroad, veterinary inspectors, serious scientists and a host of enthusiasts with a passion to see the project successfully completed. Having

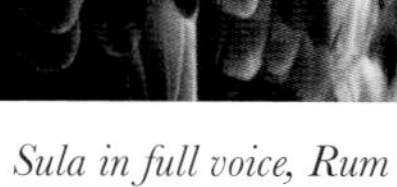

Sula in full voice, Rum

Adult Sea Eagle in flight, Canna

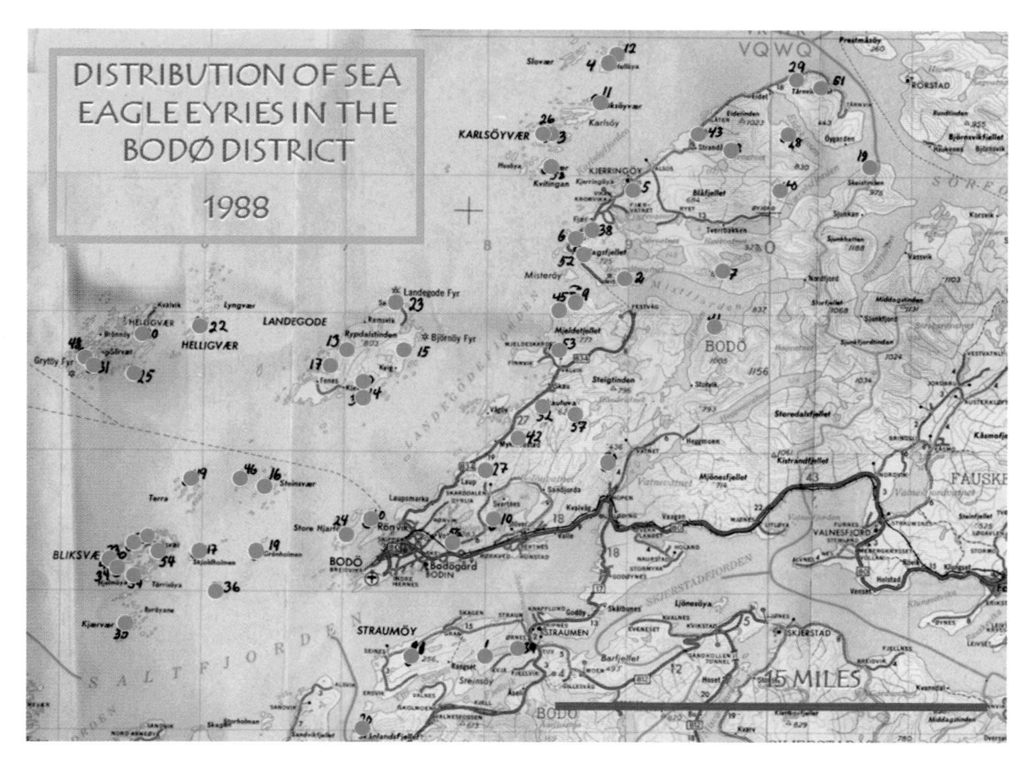

Map showing all the sea eagle nesting sites around Bodø in 1988. Others have been discovered since (Harald Misund)

started out with a modest cast of characters, the project now has hundreds of people playing a part, and at centre stage, the stars of the show, are the Sea Eagles themselves.

We had expected some to die before reaching maturity: that is the way of nature and how natural selection works for the long-term good of the species. But amongst the 319 set free in Britain and Ireland, survival has on the whole been encouragingly high. Many have matured to nest successfully and contribute progeny to the breeding population. Now, surely, it is self-perpetuating and the future of the Sea Eagle, back where it belongs, must be assured?

Our British population now stands at some about 66 pairs, having reared over 460 young wild in Scotland. With 60 young fledged, 2012 proved the best year yet and we are confident that the population will continue to go from strength to strength. And to think it all began with a casual offer of six weeks' employment on a lonely Hebridean island…

Postscript

Now, over thirty years later, I am walking along a shingle beach under the basalt cliffs of the Isle of Canna, one of the Small Isles and one of my favourite islands. Above me is Compass Hill, so named because the rock's rich iron content can upset a ship's navigational instruments. Four miles away, on the opposite side of the Sound of Canna, is the Isle of Rum, the largest of the four Small Isles. I once lived there – for nearly ten years, between 1975 and 1985 – while, on behalf of the Nature Conservancy Council, I managed the pioneering project to reintroduce White-tailed Sea Eagles back to Scotland. With no rabbits or hares in Rum some of the young, newly-released Sea Eagles quickly discovered a bunny bonanza on nearby Canna. Here there were also good seabird colonies, with boats fishing close offshore, always ready to chuck some of their catch to attract the eagles closer.

Back in Rum I had been providing food in the form of venison and fish on the eagles' cages to tide them over until they learned to hunt for themselves. I had tramped Rum's hills trying to identify the young Sea Eagles by their coloured leg rings or numbered wing tags. The Mackinnon family in Canna would inform me of any Sea Eagles there, which always gave me a fine excuse to visit, and so began my love affair with the island which I still regard as a second home.

I had first visited Rum for a week in April 1969. Only a year previously I had been on holiday in Fair Isle. This had coincided with the arrival of three young Sea Eagles from Norway (a fourth arrived later) in an unsuccessful reintroduction attempt that was never followed up. Little did I suspect that, seven years later, I would become so intimately involved in a fresh attempt on the Isle of Rum. After a hesitant start, we managed to release no fewer than 82 young Norwegian Sea Eagles over eleven seasons and further releases would take place.

Today is a calm day in early May with clear blue skies. I know I will soon have to turn back when the beach finally gives out, below the highest cliffs on the island. But just as I turn into a small gully with high precipitous sides, a Sea Eagle appears overhead. It is soon joined by its mate. Together they circle, wings spread like condors, their white tails shining in the sun like silver crescent moons. They croak in alarm, their duet approaching an excited crescendo, the smaller male's voice slightly higher pitched.

And then, as if on cue, the resident Peregrine pair tumble off the cliff and screech at the intrusion. The four birds perform above me - the lethargic giants soaring against the blue sky, the nimble little falcons cavorting angrily around them. Their chorus echoes loudly around the cliffs as I beat a hasty retreat. All the efforts have been worthwhile…

References

Armstrong, E.A. 1958. *The Folklore of Birds*. Collins, London.

Bannerman, D.A. 1956. *The Birds of the British Isles, Vol. 5*. Oliver & Boyd, London.

Banzhaf, W. 1937. *Der Seeadler. Dohrniana* 16:3–41

Baxter, E.V. and Rintoul, L.J. 1953. *The Birds of Scotland, Vol. 1*. Oliver & Boyd, London.

Bent, A.C. 1937. *Life Histories of North American Birds of Prey*. Dover, London. 1961

Bere, S.V. 1954. *The Observer's Book of Birds*. Frederick Warne, London.

Bijleveld, M.F. I.J. 1974. *Birds of Prey in Europe*. Macmillan, London.

Birkhead, T. 2012. *Bird Sense*. Bloomsbury, London.

Blackburn, H. 1895. *Birds from Moidart and Elsewhere*. David Douglas, Edinburgh.

Booth, E.T. 1881–87. *Rough Notes on the Birds Observed During Twenty Five Years' Shooting and Collecting in the British Islands*. 3 vols. Porter, London.

Brodkorb, P. 1955. Number of feathers and weights of various systems in a Bald Eagle. Wilson Bulletin 59:3–20

Brown, P. and Waterston, G. 1962. *The Return of the Osprey*. Collins, London.

Buckley, T.E. and Harvie-Brown, J.A. 1891. *A Vertebrate Fauna of the Orkney Islands*. David Douglas, Edinburgh.

Cade, T. and Temple, S.A. 1977. The Cornell University Falcon Programme in *World Conference on Birds of Prey*. Ed. Chancellor, R. D. pp 353–69. I.C.B.P., Vienna.

Christensen, J. 1979. Den gronlandske Havørns *Haliaeetus albicilla* Brehms ynglebiotop redeplacering og rede. (The breeding habitat, nest and nest of the Greenland White-tailed Eagle in Danish with an English summary) *Dansk Ornithologisk Forenings Tidsskrift* 73: 131–56.

Colquhoun, J. 1888. *The Moor and the Loch*. W. Blackwood & Sons. Edinburgh and London.

Cowles, G.S. 1969. Alleged skeleton of osprey attached to carp. *British Birds* 62:541–4

Cramp, S. and Simmons, K.E.L. 1977–1994. *Handbook of the Birds of Europe, the Middle East and North Africa*, Vol. 2. Oxford University Press, Oxford.

Darling, F. F. 1955. *West Highland Survey*. Oxford University Press.

del Hoyo, J., Elliot, A. and Sargatal, J. (Eds.). 1994. *Handbook of the Birds of the World*, Vol. 2. Lynx Edicions, Barcelona.

Dementiev, G.P. and Gladkov, H.A. 1966. *The Birds of the Soviet Union*. Jerusalem Programme for Scientific Translations.

Dennis, R.H. 1968. Sea Eagles. *Fair Isle Bird Observatory Report* 21:17–21

Dennis R.H. 1969. Sea Eagles. *Fair Isle Bird Observatory Report* 22:23–9

Dennis, R.H. 2008. *A Life of Ospreys*. Whittles Publishing, Dunbeath.

Deppe, H.J. 1972. Einige Verhaltenbeobachtungen in einem Doppelhorst von Seeadler (*Haliaeetus albicilla*) und Wanderfalke (*Falco peregrinus*) in Mecklenburg. *Journal of Ornithology* 113:440–4

Dixon, C. 1900. *Among the Birds in Northern Shires*. London.

D'Urban, W.S.M. and Matthew, Rev. M.A. 1892. *The Birds of Devon*. Porter, London.

Edmondston, T. 1844. A Fauna of Shetland . *Zoologist* 459–67, 551–52.

Ferguson-Lees, J. and Christie, D.A. 2001. *Raptors of the World*. Christopher Helm, London.

Fergusson, C. 1885. The Gaelic names of Birds, Part 1. *Transactions of the Gaelic Society of Inverness* 11:240–60.

Fischer, W.1979. *Die Seeadler*. A.Ziemsen Verlag, Wittenberg-Lutherstadt.

Fisher, J.1966. *The Shell Bird Book*. Edbury Press and Michael Joseph, London.

Flerov, A.I. 1970 [On the ecology of the White-tailed Eagle in the Kandalaksha Bay] *Proceedings of the Kandalaksha State Nature Reserve (Murmansk)*, No. VII: 215–32 (in Russian)

Folkestad, A.O.2003. Status of the White-tailed Sea Eagle in Norway. In Helander B. *et al.* (Eds.) *Sea Eagle 2000.* Proceedings from an International Conference in Bjørkø, Sweden, 13–17 September 2000. SNF, Stockholm.

Folkestad, A.O.2003. Nest site selection and reproduction in the White-tailed Sea Eagle in Møre og Romsdal county, western Norway, in relation to human activity. In Helander B. *et al.* (Eds.) *Sea Eagle 2000.* Proceedings from an International Conference in Bjørkø, Sweden, 13–17 September 2000. SNF, Stockholm.

Friedman, H. 1950. The Birds of North and Middle America. 6 volumes. *US National Museum Bulletin* 60, 793 pp.

Glutz von Blotzheim,U.,Bauer, K. and Bezzel, E. 1971. *Handbuch der Vögel Mitteleuropas.* Band 3. Akademische Verlagsgellschaft, Frankfurt.

Gray, R.1871. *The Birds of the West of Scotland, including the Outer Hebrides.* Glasgow.

Green, R.E., Pienkowski, M.W. and Love, J.A. 1996. Long-term viability of the reintroduced population of the white-tailed eagle *Haliaeetus albicilla* in Scotland. *Journal of Applied Ecology* 33:357–68

Hall, J. 1976. The Golden Eagle and sheep stocks in North Uist. NCC unpublished report.

Hario,M. 1981. Vinterutfodring av Havsörn i Finland. [Winter feeding of White-tailed Eagles in Finland] In *Projekt Havsörn I Finland och Sverige.* Ed. T Stjernberg, pp 113–122. Jord-och Skogsbruksministeriet, Helsinki. (In Finnish with English summary).

Harrison, C.J.O. and Walker, C.A. 1973. An undescribed extinct fish eagle from the Chatham Islands. *Ibis* 115(2):274–277

Harvie-Brown, J.A. 1906. *A Fauna of the Tay Basin and Strathmore.* David Douglas, Ediinburgh.

Harvie-Brown, J.A. and Buckley, T.E. 1887. *A Vertebrate Fauna of Sutherland, Caithness and West Cromarty.* David Douglas, Edinburgh.

Harvie-Brown, J.A. and Buckley, T.E. 1888. *A Vertebrate Fauna of the Outer Hebrides.* David Douglas, Edinburgh.

Harvie-Brown, J.A. and Buckley, T.E. 1892. *A Vertebrate Fauna of Argyll and the Inner Hebrides.* David Douglas, Edinburgh.

Harvie-Brown, J.A. and Buckley, T.E. 1895. *A Vertebrate Fauna of the Moray Basin*, 2 volumes. David Douglas, Edinburgh.

Harvie-Brown, J.A. and Macpherson, Rev. H.A. 1904. *A Vertebrate Fauna of the North-west Highlands and Skye.* David Douglas, Edinburgh.

Hedges, J.W. 1984. *Tomb of the Eagles.* John Murray, London.

Helander, B. 1975. *Havsornen I Sverige.* Svenska Naturskyddsföreningen. Bokusläningen, Uddevalla, Sweden. (In Swedish with an English summary).

Helander, B. 1981. Utfodring av Havsorn I Sverige. [Feeding White-tailed Eagles in Sweden). In *Projekt Havsörn I Finland och Sverige.* Ed. T. Stjernberg pp. 113–122. Jord-och Skogsbruksministeriet, Helsinki. (In Swedish with English summary).

Helander, B., Marquiss, M.and Bowerman, W. (Eds) 2003. Sea Eagle 2000. Proceedings from an international conference at Björkö, Sweden, 13–17 September 2000. Swedish Society for Nature Conservation/SNF & Åtta..45 Tryckeri AB. Stockholm.

Houston, D.1977. The effect of Hooded Crows on hill sheep farming in Argyll, Scotland: Hooded Crow damage to hill sheep. Journal of Applied Ecology 14:17–19.

Hume, J. P.and Walters, M. 2012. *Extinct Birds.* T & A.D.Poyser, London.

Kampp, K.and Wille, F. 1979. Fødevaner hos den Grønlandske Havorn *Haliaeetus albicilla groenlandicus* Brehm [Food habits of the Greenland White-tailed Eagle]. Dansk ornithologisk Forenings tidsskrift 73:157–64 (In Danish with English summary).

Leitch, A.F. 1986. Eagle predation of lambs in the Glenelg area in 1986. Unpublished report to the Department of Agriculture and Fisheries for Scotland, Edinburgh.

Lilford Lord 1885. *Coloured Figures of the Birds of the British Islands*. Vol.1. Lightfoot, London.

Lockie, J.D. 1964. The breeding density of the Golden Eagle and Fox in relation to food supply in Wester Ross, Scotland. *Scottish Naturalist* 71:67–77.

Lockie, J.D. and Stephen, D.1959. Eagles, lambs and land management on Lewis. *Journal of Animal Ecology* 28:43–50.

Lodge, G.1946. *Memoirs of an Artist Naturalist*. Gurney & Jackson, London.

Love, J.A.1977. The reintroduction of the Sea Eagle to the Isle of Rhum. *Hawk Trust Annual Report* 1977:16–18.

Love, J.A.1979. The daily food intake of captive White-tailed Eagles. *Bird Study* 26:64–66.

Love, J.A.1980. White-tailed Eagle reintroduction on the Isle of Rhum. *Scottish Birds* 11:65–73.

Love, J.A. 1982. Harvie-Brown: a profile. *Scottish Birds* 12:49–53.

Love, J.A. 1983. *The Return of the Sea Eagle*. Cambridge University Press.

Love, J.A. 1988. *The re-introduction of the White-tailed Sea Eagle to Scotland 1975–87*. NCC Research and Survey in Nature Conservation Report 12. 48 pp. Nature Conservancy Council, Peterborough.

Love, J.A. 2001. *Rum: a landscape without figures*. Birlinn, Edinburgh.

Love, J.A. 2005. A Sad Saga of Special Sea Eagles. *Teachd an Tir* Spring 2005, p 1.

Love, J.A. 2008. First and last of Britain's other eagle. *Scottish Bird News* 89: 9.

Love, J.A. 2009. *A Natural History of St. Kilda*. Birlinn, Edinburgh.

Low, Rev. G. 1813. *A Tour Through the Islands of Orkney and Shetland in 1774*. Edinburgh. (Melven, Inverness, 1978 reprint).

Lysaght, S. 2004. *Science and Irish Culture* (Eds. Attis, D. and Mollan, C.) RDS, Dublin.

Macgillivray, W. 1836. The Birds of the Outer Hebrides. Edinburgh Journal of Natural and Geographical Science 142:245–50; 401–11; 87–95; 161–5; 321–34.

Macgillivray, W. 1886. *Descriptions of the rapacious birds of Great Britain*. Edinburgh.

Mackenzie, O.H. 1928. *A Hundred years in the Highlands*. Geoffrey Bles, London.

Maclennan, A.and Evans, R. 2003. Public viewing of White-tailed Sea Eagles – take the birds to the people or the people to the birds? pp 417-22 In Helander, B., Marquiss, M. and Bowerman, W. (Eds) *Sea Eagle 2000*. Proceedings from an International Conference in Bjørkø, Sweden, 13–17 September 2000. SNF, Stockholm.

Macmillan, A.1968. Editorial: Sea Eagles at Fair Isle. *Scottish Birds* 5:121–2.

Macpherson, Rev. H.A. 1892. *A Vertebrate Fauna of Lakeland*. David Douglas, Edinburgh.

Macpherson, Rev. H.A. 1897. *A History of Fowling*. David Douglas, Edinburgh.

Martin, M. 1697. *A Late Voyage to St. Kilda*. London. (Reprint 1986. Mercat Press, Edinburgh).

Martin, M. 1716. *Description of the Western Isles of Scotland*. Andrew Bell, London. (Reprint 1976. James Thin, Edinburgh).

Maxwell, Sir H. 1907. *Memories of the Month*, first series. Edward Arnold, London.

McMillan, R. L. 2005. *Skye Birds*. skyebirds.com, Broadford.

Meinertzhagen, R. 1959. *Pirates and Predators*. Oliver & Boyd, Edinburgh.

Miller Mundy, A. 1996. Riding in the Eagle's wake. *The Field*.

Mitchell, W.R.and Robson, R.W. 1976. *Lakeland Birds: a visitor's handbook*. Dalesman Books, Clapham.

Morris, Rev. F.A.O. 1851. *A History of British Birds*, Vol.1. London.

Mynott, J. 2009. *Birdscapes*. Princeton University Press, London and Princeton.

Newton, A.1860. Memoir of the late John Wolley Jun. *Ibis* 2:172–185.

Newton, I.1979. Population Ecology of Raptors. T & A.D. Poyser, Berkhamsted.

New Statistical Account of Scotland. 1835–45. 15 vols. Edinburgh.

Nye, P.E. 1982 A biological and economic review of the hacking process for the restoration of Bald Eagles. Presented at the International Bald Eagle/Osprey Symposium, October 28–29 1981, Montreal, Canada.) *In Federal Aid to endangered species. New York Project E-1, Performance Report.* 16 pp.

Oehme, G. 1961. Der Bestandsentwicklung des Seeadlers in Deutschland mit Untersuchengen zur Wahl der Bruthbiotopen, pp.1–61. *Beiträge zur Kenntnis deutscher Vögel.* Ed. H. Schildmacher. Jena.

Oehme, G. 1977. Seeadler, pp. 134–135. *Die Vogelwelt Mecklenburgs.* Eds. G. Klafs and J. Stubbs. Fischer, Jena.

Old Statistical Account 1791–99. 21 vols. Edinburgh.

Pain, S. 1988. Vagrant Eagle 'recrosses' the Atlantic. *New Scientist,* 25 February, p.28.

Pennant, T. 1774. *British Zoology.* London.

Pennington, M., Osborn, K., Harvey, P., Riddington, R., Okill, D., Ellis, P. and Huebeck, M. 2004. *The Birds of Shetland.* Christopher Helm. London.

Ritchie, J. 1920. *The Influence of Man on Animal Life in Scotland.* Cambridge University Press.

Rudebeck, G.1951. The choice of prey and modes of hunting of predatory birds with special reference to their selective effect. *Oikos* 3:200–231.

Salomonsen, F.1950. *Grönlands Fugle.* Rhodos, Copenhagen.

Sandeman, P.1957. The breeding success of golden eagles in the southern Grampians. *Scottish Naturalist* 69:148–152.

Sandeman, P.1965. Attempted reintroduction of White-tailed Eagle to Scotland. *Scottish Birds* 3:411–412.

Scott, Sir W.1998. *The Voyage of the Pharos.* Scottish Library Association, Hamilton.

Schnurre, O.1956. Ernä hrungsbiologische Studien an Raubvögeln und Euten der Darsshalbinsel (Mecklenburg). *Beiträge Vogelkunde* 4:211–245.

Seebohm, H.1883. *A History of British Birds.* Vol.1. London.

Sim, G. 1903. *The Vertebrate Fauna of Dee.* Aberdeen.

Simms, I.C., Ormston, C.M., Somerwill, K.E., Cairns, C.L., Tobin, F.R., Judge. J. and Tomlinson, A. 2010. *Wester Ross Lamb Cohort Study 2009: Final Report.*

Scottish Natural Heritage Commissioned Report No.27291.

Sexton, D.1988. Irish Bald Eagle. *BBC Wildlife* Vol. 6 (3).

Southern, W.E. 1964. Additional observations on winter Bald Eagle populations: including remarks on biotelemetry techniques and immature plumages. *Auk* 76:121–137.

Stjerberg, T. 1981. Projekt Havsörn I Finland. In *Projekt Havsörn I Finland och Sverige.* Ed. T Stjernberg, pp. 31–60 Jord-och Skogsbruksministeriet, Helsinki. (In Finnish with English summary).

St. John, C. 1849. *A Tour of Sutherland.* David Douglas, Edinburgh.

Tingay, R.E. 2005. *Historical distribution, contemporary status and co-operative breeding in the Madagascar Fish Eagle: implications for conservation.* PhD thesis, University of Nottingham.

Tulloch, J. 1904. Albino Sea Eagle in Yell, Shetland. *Annals of Scottish Natural History* 52:245.

Tulloch, R.J. 1978. The eagle and the baby. *Scots Magazine* 109:260–4.

Ussher, R.J. and Warren, R. 1900. *The Birds of Ireland.* Gurney and Jackson, London.

Uttendörfer, O.1939. *Die Ernährung der Deutschen Raubvogel und Eulen.* Neumann, Neudamm.

Venables, L.S.V. and Venables, U.M. 1955. *The Birds and Mammals of Shetland.* Oliver & Boyd, Edinburgh.

Wagner, F.H. 1972. Quoted in Eagles and sheep: a viewpoint by E. G. Bolen, *Journal of Range Management* 28:11–17.

Waterston, G. 1964. Sea Eagles in Norway. *Bird Notes* 31:18–23.

Waterston, G. 1964. Studies of less familiar birds: 130 White–tailed Eagle. *British Birds* 57:458–466.

Waterston, G. 1968. Sea Eagles for Fair Isle *Birds* 2:111–112.

Watson, J. 1979 and 2010 (first and second editions). *The Golden Eagle*. T & A.D. Poyser, London.

Watson, J., Leitch, A.F. and Broad, R.A.1992. The diet of the sea eagle *Haliaeetus albicilla* and Golden Eagle *Aquila chrysaetos* in western Scotland. *Ibis* 134:27–31.

Weir, D.N. 1973. A case of lamb-killing by golden eagles. *Scottish Birds* 7:293–301.

Weir, T. 1982. My month. *Scots Magazine* 118:72–76.

Wille, F. 1979. Den Grønlandske Havorns *Haliaeetus albicilla groenlandicus* Brehm fødevalg – metode og foreløbige resultater. [Choice of food of the Greenland White-tailed Eagle – method and preliminary results] *Dansk ornithologisk forenings Tidsskrift* 73:165–170. (In Danish with English summary).

Willgohs, J.F. 1961. *The White-tailed Eagle Haliaeetus albicilla albicilla (L.) in Norway*. Arbok for Universitetet, Bergen.

Williamson, K. 1958. *The Atlantic Islands*. Oliver & Boyd. (Second edition 1970, Routledge, London and Kegan Paul).

Wilson, K.-J. 2004. *Flight of the Huia: ecology and conservation of New Zealand's frogs, reptiles, birds and mammals*. Canterbury University Press, New Zealand.

Wolley, J. 1902. *Ootheca Wolleyana* vol.1 (Ed.) A. Newton, R.H. Porter, London.

Worthy, T.H. and Holdaway, R.N. 2002. *The Lost World of the Moa*. Indiana University Press, Bloomington and Indianapolis.

Yalden, D.W. and Albarella, U. 2009. *The History of British Birds*. Oxford University Press.

Yarrell, W. 1871. *A History of British Birds*, vol.1, Van Voorst, London.

Further Reading

Bainbridge, I.P., Evans, R.J., Broad, R.A., Crooke, C.H., Duffy, K., Green, R.E., Love, J.A. and Mudge, G.P. 2003. Re-introduction of white-tailed eagles (*Halieetus albicilla*) to Scotland, pp. 393–406. In Thompson D.B.A., Redpath, S.M., Fielding, A.H., Marquiss, M. and Galbraith, C.A. (Eds.) *Birds of Prey in a Changing Environment*. SNH, HMSO.

Colquhoun, J. 1862. A Skye-Lark. *Blackwood's Edinburgh Magazine* 92:151–162.

Dawson, E.W. 1961. An extinct eagle in the Chatham Islands. *Notornis* 9:171–172.

Dennis, R.H. 2003. A re-appraisal of the future range of the White-tailed Eagle in the British Isles and western Europe and proposals for an enhanced recovery scheme, pp. 413–416. In Helander, B., Marquiss, M. and Bowerman, W. (Eds.) *Sea Eagle 2000*. Proceedings from an International Conference in Bjørkø, Sweden, 13–17 September 2000. SNF, Stockholm.

Evans, I.M., Love, J.A., Galbraith, C.A. and Pienkowski, M.W. 1994. Population and range restoration of threatened raptors in the United Kingdom. In *Raptor Conservation Today* (Eds. Meyburg, B.-U. and Chancellor, R. D.) pp. 447–457. World Working Group on Birds of Prey, Berlin.

Evans, R., Broad, R., Duffy, K., Maclennan, A., Bainbridge, I. and Mudge, G. 2003. Re-establishment of a breeding population of White-tailed Eagles in Scotland. 397–403. In Helander, B., Marquiss, M. and Bowerman, W. (Eds) *Sea Eagle 2000*. Proceedings from an International Conference in Bjørkø, Sweden, 13–17 September 2000. SNF, Stockholm.

Evans, R.J., Wilson, J.D., Amar, A., Douse, A., Maclennan, A., Ratcliffe, N., and Whitfield, D. P. 2009. Growth and demography of a reintroduced population of White-tailed Eagles *Haliaeetus albicilla*. Ibis 151:244–254.

Evans, R.J., Pearce-Higgins, J., Whitfield, D.P., Grant, J., Maclennan, A.and Reid, R. 2010. Comparative nest habitat characteristics of sympatric White-tailed *Haliaeetus albicilla* and Golden eagles *Aquila chrysaetos* in western Scotland. *Bird Study* 57:473–482.

Evans, R.J., O'Toole, L. and Whitfield, D.P. 2012. The history of eagles in Britain and Ireland: an ecological review of place name and documentary evidence from the last 1500 years. *Bird Study* 59:335–349.

Halley, D.J. 1998. Golden and white-tailed eagles in Scotland and Norway: coexistence, competition and environmental degradation. *British Birds* 91:171–179.

Halley, D.J. and Gershaug, J.O. 1998. Inter- and intra-specific dominance relationships and feeding behaviour of golden eagles *Aquila chrysaetos* and Sea Eagles *Haliaeetus albicilla* at carcasses. *Ibis* 140:295–301.

Love, J.A. 1989. *Eagles*. Whittet Books.

Love, J.A. 2003. *A history of the Sea Eagle in Britain*, pp. 39–50. In Helander, B.,Marquiss, M. and Bowerman, W. (Eds.) *Sea Eagle 2000*. Proceedings from an International Conference in Bjørkø, Sweden, 13–17 September 2000. SNF, Stockholm.

Love, J.A. 2007. *White-tailed Eagle*, pp. 451–455. In Forrester, R.W., Andrews, I.J., McInerny, C.J., Murray, R.D., McGowan, R.Y., Zonfrillo, B., Betts, M.W., Jardine, D.C. and Grundy, D.S. (Eds.) *The Birds of Scotland*, 2 vols. The Scottish Ornithologists' Club, Aberlady.

Love, J.A. 2006. *Sea Eagles*. 35 pp. Naturally Scottish series, Scottish Natural Heritage, Redgorton.

Love, J.A. 2010. *White-tailed eagles, Scotland*, pp. 201–206. In Tingay, R.E. and Katzner, T.E. (Eds.) *The Eagle Watchers*. Cornell University Press, Ithaca and London.

Love, J.A. and Ball, M.E. 1979. White-tailed sea eagle (*Haliaeetus albicilla*) Reintroduction to the Isle of Rhum, Scotland 1975–1977. *Biological Conservation* 16:23–30.

Love, J.A., Ball, M.E. and Newton, I. 1978. White-tailed Eagles in Britain and Norway. *British Birds* 71:475–481.

Madders, M. and Marquiss, M. 2003 A comparison of the diet of White-tailed eagles and Golden eagles breeding in adjacent ranges in west Scotland, pp 289–295. In Helander B. et al. (Eds) *Sea Eagle 2000*. Proceedings from an International Conference in Bjørkø, Sweden, 13–17 September 2000. SNF, Stockholm.

Marquiss, M., Madders, M. and Carss, D.N. 2003. White-tailed eagles (*Haliaeetus albicilla*) and Lambs (*Ovis aries*), pp. 471–479. In Thompson, D.B.A., Redpath, S.M., Fielding, A.H., Marquiss, M. and Galbraith, C.A. (Eds) *Birds of Prey in a Changing Environment*. SNH, HMSO.

Misund, H. 1999. Ornestammen I Skottland. *Havørna* 10:16–17.

Misund, H. 2001. Iolair na mara (Havørna). *Havørna* 12:43–45.

Misund, H. 2005. Vekst av Havørnstammen I Skottland. *Havørna* 16:30–34.

Nellist, K. and Crane, K. 1999. *Island Eagles*. Cartwheeling Press, Glenbrittle.

Royal Society for the Protection of Birds. (No date).*White-tailed Eagles*. RSPB, Sandy.

Sexton, D. 1987. The Sea Eagle's Return. *Birds*. August: 22–24.

Whitfield, D.P., Evans, R. J., Broad, R.A., Haworth, P.F., Madders, M. and McLeod, D.R.A. 2002. Are reintroduced white-tailed eagles in competition with golden eagles? *Scottish Birds* 23:36–45.

Whitfield, D.P., Douse, A., Evans, R.J., Grant, J., Love, J.A., McLeod, D.R.A., Reid, R.and Wilson, J. D. 2009. Natal and breeding dispersal in a reintroduced population of White-tailed Eagles *Haliaeetus albicilla*. *Bird Study* 56:177–186.

Wille, F. and Kampp, K. 1983. Food of white-tailed eagle *Haliaeetus albicilla* in Greenland, *Holarctic Ecology* 6:81–88.

Yalden, D.W. The older history of the White-tailed Eagle in Britain. *British Birds* 100:471–480.

ndix I

White-tailed sea eagle in different languages

Albanian: Shqiponja e detit
Arabic: عقاب ابيض الذنب, عقاب ابيض الذنب(عقاب البحر), عقاب البحر
Armenian: [Spitakapoch Artsiv], Սպիտակապոչ Արծիվ
Azerbaijani: Ağquyruq dəniz qartalı
Basque: Àguila marina, Itsas arrano buztanzuria
Belarusian: Арлан-белахвост
Breton: Ar morerer lost gwenn
Bulgarian: Белоопашат морски орел, Морски орел
Catalan: Àguila marina
Chinese: [bai-wei diao], [bai-wei hai-diao], [bai-wei jiu], [jie-bai diao], [zhima diao], 洁白鵰, 白尾海雕, 白尾海鵰, 白尾鵰, 白尾鷲, 芝麻鵰
Chinese (Taiwan): [bai-wei jiu], 白尾鷲
Cornish: Er an mor
Croatian: Orao štekavac, Štekavac
Czech: orel moøský, Orel morský, Orel mořský
Danish: Havørn
Dutch: Zeearend
English: Cinereous Eagle, Erne or Earn, Fish Hawk or Fish Eagle, European Sea Eagle, European Sea-eagle, Gray Eagle, Gray Sea Eagle, Gray Sea-eagle, Grey Eagle, Grey Sea Eagle, Grey Sea-eagle, White-tailed Eagle, White-tailed Fish-eagle, White-tailed Sea Eagle, White-tailed Sea-eagle
Esperanto: Blankvosta maraglo
Estonian: Merikotkas
Faroese: Havørn, Ørn
Finnish: Merikotka
French: Milan royal, Pygargue à queue blanche
Gaelic: Iolair Sùil na Grèine, Iolair-mhara, Iolar Bhuidhe, Iolar Riamhach etc
Galician: Àguila marina, Pigargo
Georgian: ანუ ფსოვი, თეთრკუდა არწივი
German: Seeadler
Greek: (Ευρωπαϊκός) Θαλασσαετός, Θαλασσαετός
Greenlandic: Nattoralik
Hebrew: עיטם לבן זנב, עיטם לבן־זנב, עיטם לבן-זנב, עיטם לבן זנב
Hungarian: Rétisas
Icelandic: Haförn
Irish: Iolair mara, Iolar mara
Italian: Aquila di mare, Aquila di mare codabianca
Japanese: ojirowashi, Ojiro-washi
Japanese: オジロワシ
Kazakh: Аққұйрықты субүркіті
Khakas: Хузургул, Хуучын
Korean: 흰꼬리수리
Latin: Haliaeetus albicilla

Latvian: Jūras ērglis
Lithuanian: Jurinis erelis, Jūrinis erelis
Macedonian: Белоопашест орел
Maltese: Ajkla tad-Denb Abajd
Manx: Urley Erne, Urley marrey
Mongolian: Усны цагаан суулт бургэд
Norwegian: Havørn, Klekse
Polish: bielik, bielik (zwyczajny), Bielik zwyczajny
Portuguese: àguia rabalva, Águia-rabalva
Romansh: Evla da mar
Russian: Orlan-belokhvost, белохвостый орлан, Орлан-белохвост, Орлан-долгохвост
Sami: Mearragoaskin
Scots Gaelic: Iolair mhara, Iolar bhuidhe, Iolar riamhach
Serbian: Orao belorepan, orao bjelorepan, Орао белорепан Swedish: Havsörn
Slovak: Orliak morský, Orol morský
Slovenian: Belorèpec, orel belorepec
Spanish: Pigargo, Pigargo coliblanco, Pigargo Europeo
Swedish: Havsörn
Thai: **นกอินทรีหางขาว**
Turkish: Akkuyruklu kartal, Ak-kuyruklu Kartal, Beyaz-kuyruklu Kartal
Tuvinian: Ак кудуруктуг орлан
Ukrainian: Орлан-білохвіст, Орлан-білохвіст , Орлан-довгохвіст, Сіруватень
Welsh: Eryr cynffon wen, Eryr gynffonwen, Eryr tinwen, Eryr tinwyn, Eryr y môr, Mor eryr

Appendix II

Current known status of White-tailed sea eagles (numerous sources)

EUROPE

ALBANIA	0	Extinct c.1975
AUSTRIA	7–10	Since 1971
BELARUS	100–150	Increasing
BOSNIA HERZEGOVINA	5	?
BULGARIA	10–15	Stable
CROATIA	150	Increasing
CZECH REPUBLIC	60	Increasing
DENMARK	7	Increasing since return in 1954
EGYPT	0	Extinct 19th century
ESTONIA	80–90	Increasing
FINLAND	250	Increasing
FRANCE	0	Extinct Corsica since 1940s
GERMANY	c. 600	Increasing
GREECE	6	Decreasing
GREENLAND	150–170	Stable
HOLLAND	3	Since 2006
HUNGARY	204–210	Increasing
ICELAND	66	Increasing
IRELAND	0	Extinct c.1910; reintroduced 2007
ITALY	0	Extinct Sardinia c.1960
LATVIA	20–25	Increasing
LITHUANIA	65	Increasing
MACEDONIA	0–2	Decreasing
MOLDOVA	0	Extinct c.1975
MONTENEGRO	0	?
NORWAY	1900–2200	Increasing
POLAND	500	Increasing
ROMANIA	30	Decreased now stable?
RUSSIA	2,000	Increasing
SERBIA	70–100	Increasing
SLOVAKIA	4–8	Increasing
SLOVENIA	7–11	Increasing
SWEDEN	350	Increasing
UKRAINE	80–100	Increasing
UNITED KINGDOM	57	Increasing
EUROPE TOTAL	6,538–7,222+ pairs	

ASIA

ARMENIA	0	Extinct c.1960
AZERBAIJAN	7–8	Decreasing
CHINA	Rare	?
GEORGIA	0	Extinct 1960s?
IRAN	A few pairs	In 1980s
IRAQ	0	Extinct?
ISRAEL	1?	Extinct early 1950s; reintroduced 1998/2007
JAPAN	39	In 1990–92
KAZAKHSTAN	95–110?	Decreasing?
MONGOLIA	50–60	Decreasing
RUSSIA	2,000–7,000?	?
SYRIA	0	Extinct
TURKEY	10–15	Decreasing
TURKMANISTAN	0	Extinct?
UZEBEKISTAN	Very rare	Decreasing
ASIA TOTAL	2,202–2,233+ pairs	
WORLD TOTAL	8,740–9,455+ pairs	

Appendix III

Numbers of Sea Eagles released in Scotland (Phases I–III) and Ireland

Year		Male	Female	Unknown	Total			
Phase I								
1975		0	3	0	3			
1976		5	4	0	9			
1977		2	2	0	4			
1978		4	3	0	7			
1979		3	3	0	6			
1980		5	3	0	8			
1981		3	2	0	5			
1982		3	7	0	10			
1983		3	7	0	10			
1984		6	4	0	10			
1985		5	5	0	10			
Total		39	43	0	82			
Phase II								
1993		4	6	0	10			
1994		5	5	0	10			
1995		1	4	1	6			
1996		3	4	3	10			
1997		5	4	1	10			
1998		7	5	0	12			
Total		25	28	5	58			
						Ireland		
Phase III						male	female	total
2007		9	6	0	15	8	7	15
2008		8	7	0	15	12	8	20
2009		7	7	0	14	7	13	20
2010		12	7	0	19	16	6	22
2011		10	6	0	16	15	8	23
2012		2	4	0	6	0	0	0
Totals								
Phase I		39	43	0	82			
Phase II		25	28	5	58			
Phase III		48	37	0	85			
Phases I-III		112	108	5	225	58	42	100

Appendix IV

Establishment in the wild of reintroduced Sea Eagle population

Year	Areas occupied	Territorial pairs	Clutches laid	Broods hatched	Broods successful	Young fledged	Young fledged per breeding pair	Young fledged per territorial pair
1983	3	3	3*	0	0	0		
1984	5	5	2	0	0	0		
1985	7	7	4	1	1	1	1.00	0.14
1986	7	7	5	1	1	2	2.00	0.29
1987	9	9	6	2	2	3	1.50	0.33
1988	9	9	6	1	1	2	2.00	0.22
1989	9	9	6	3	3	5	1.67	0.56
1990	12	9	8	2	2	2	1.00	0.22
1991	10	9	7	4	4	7	1.75	0.78
1992	10	8	7	4	4	7	1.00	0.88
1993	10	8	6	4	4	5	1.25	0.63
1994	11	10	8	4	4	5	0.63	0.50
1995	14	11	10	5	5	7	0.70	0.64
1996	14	12	12	8	7	9	0.75	0.75
1997	14	14	11	6	5	9	0.64	0.64
1998	19	19	16	9	9	13	0.81	0.68
1999	20	20	16	9	6	11	0.69	0.55
2000	23	22	19	12	8	12	0.63	0.55
2001	24	23	17	10	7	11	0.65	0.48
2002	26	25	22	14	8	12	0.55	0.48
2003	31	31	25	20	16	26	1.04	0.84
2004	32	32	28	19	15	19	0.68	0.59
2005	33	33	28	21	17	24	0.86	0.73
2006	36	36	31	25	21	29	0.94	0.81
2007	42	42	35	31	24	34	0.97	0.81
2008	44	44	35	21	20	28	0.85	0.64
2009	46	46	39	31	24	36	0.92	0.78
2010	51	51	45+	33+	33+	45	1.00	0.73
2011	58–60	52–57	45–52	31+	31+	43	1.05	0.72
2012	66–72	66	58+	53+	41	60	0.96	0.68
Total						467		

Index

African fish eagle 16

Bald Eagle 17, 25, 31, 200
    restocking 100
BBC TV coverage 128, 134, 137, 209, 226
Brahminy kite 13
BTO ring 131

cages 109, 125, 129, 204
Canna 180, 190
classification 14

data recording 47
decline in numbers in 19th century 82
    human population and sheep numbers 86
    reasons 84
    shooting 86–8
diet 22–3, 46–58, 159
    birds 52–4
    collecting information about 46–7
    fish 49–52
    fresh-water prey 51
    lamb killing, claims of 214–25
    mammals 55–6
    pellets as indicators 46–7
    variation with season 48

early reintroduction 1–11
eggs 21, 89, 93
    weight 173
Egyptian vulture 13
extinction 83
eyrie 19, 20–2, 114, 166, 179, 188
    Golden Eagle 90

Fair Isle Bird Observatory 7, 9
    early stages 6
Fair Isle early releases 9–10, 109
falconry 32
Faroe 64
food and feeding 44–58

Gaelic names xi
Golden Eagle 13
    clutches stolen 93
    eyries 90
    persecution 93–4

Haast's eagle 18, 99

history, legends and symbolism 28–43
    Anglo-Saxon 63
    heraldry 29
    mythical connections 40
    pub names 41
    reference in battles 63
    Sea Eagle and children 41–3
    stamps 30
    trapping 38
human population and sheep numbers 86

introductions, of alien species 99–100
ITN TV coverage 128

lamb killing, claims of 95, 214–25
Lewis 167, 170, 171
Loch Garten 224

Madagascar Fish Eagle 22
    'Cut Off' 36, 37
Mull 75, 163–5, 166
    Eagle Watch 225
    first chick 167
    Sea Eagle breeding attempts 165

nests 20–1, 118, 166
    building 166
New Zealand
    bone finds 14
    Haast's Eagle 18, 99
    introduced species 99
Norway
    Bodø 111–12
    census of breeding Sea Eagles 110
    eaglets from 121, 129, 135, 143, 145, 154, 206
    Helligvaer 113–14
    Lofoten Islands 37
    monitoring Sea Eagles in Bodø Kommune 119
    Sea Eagle distribution 111
    Smøla archipelago 111
    source of birds 106

ørnehus, for catching Sea Eagles 37–8, 39

Pallas's Sea Eagle 15
Palm-nut vulture 13
patagial tags 132, 155–6
persecution 82–98
    clutches stolen 93

eating poisoned carrion 95, 104
extinction 83
lamb killing, claims of 95, 214–25
shooting 103–4
lack of justification 103
linked to loss of livestock 95
poisoning 7, 88, 95, 134, 155, 197, 201

reintroduction 1–11, 99–120
assessment for 100
cages 109 *see also* separate entry
choice of location 107–8
compared to introduction 100
early 1–11
feeding the eaglets 125–6
first arrival 109
guidelines for 102
options for implementing 105
population 245
Rum 11, 101 (*see separate entry*)
support from Norway 106 *q.v.*
use of radio transmitters 145–6
Reintroduction Project 101
Phase II 183 *et seq.*, 190
Phase III 203 *et seq.*
progress 158
releases 9–10, 134, 136, 137, 138, 140 *et seq.*, 149, 151, 154
and adapting to the wild 140–60
restocking 100
ringing 131
Rum 11, 101, 107, 121–39
arrival of eaglets from Norway 121, 129, 143, 145, 154, 206
eaglets prior to releases 124–6
first chick 167, 168–9
first eggs 156
first release 134, 142
food intake of captive birds 127
hand feeding 130
patagial tags 132, 155–6
Reintroduction Project 101
releases 9–10, 134, 136, 137, 138, 140 *et seq.*, 149, 151, 154
releases and adapting to the wild 140–60
relief map 102
ringing 131
start of reintroduction 101
tethered birds 141–5
tether site 147

Sanford's Sea Eagle 30

Sea Eagle
archaeological finds 59–62
biology 19
breeding successes 210–1
brood size 177, 178
chick weight and development 174–7
clutch size 21–2, 173, 178
diet 22–3, 46–58 *q.v.*
distribution in 19th century 91
eggs 21, 89, 93
eyrie 20–2
food and feeding 44–58
history, legends and symbolism *see separate entry*
incubation 22, 173
last in Britain 25
legal protection 104
observations on breeding 172
pellets 46–7, 128
pesticide levels in eggs 171
plumage 18, 26
prey 22–3 *see also* diet
ranges 18, 24
sightings in England 103
sizes 18
species 16
tourist appeal 223, 228
trapping 38
vision 44–5
weights 24
wing length 17, 24
Sea Eagles in Britain, historical record 59–81
distribution in 19th century 91
England 103
Lake District 66–7
prehistory 59–62
Scotland 68 *q.v.*
St. Kilda 65
Sea Eagles in Greenland 47
Sea Eagles in Ireland 67, 97, 198–202
historical record 67, 97
numbers released 244
Southwest 198
Sea Eagles in Norway 105, 110–11, 113–16
monitoring by Harald Misund 119–20
Sea Eagles in Scotland, historical record 68
Ailsa Craig 68
Argyll 71
Caithness 72–3, 134
Isle of Lewis 75
Moray Firth 70
Mull 73–4

numbers released 244
Orkney 76–7
Rannoch 71
Shetland 77–80
Shiant Isles 75–6
Skye 75, 163, 171, 227
St. Kilda 65
Sutherland 72
Wester Ross 72, 183
Skye 75, 163, 171, 229
Southwest Ireland 198
species 12–16
Steller's Sea Eagle 17, 18

taxonomy xii, 12, 14
terminology x
TV coverage 128, 134, 137, 209, 226

Uist 190

Vulturine Fish Eagle 12

weights 24
of eggs 173
Wester Ross 72, 183
Shieldaig 227
White-bellied Sea Eagle 16
wing length 24

Index of names

Ball, Martin 108, 112, 154
Boyd, (John) Morton 2–3, 132
Broad, Roger 135, 163
Christiansen, Hallstein 115
Colquhoun, John 87, 96
Cook, Mike 170, 174
Coombes, Richard 136
Crane, Ken 167, 212–3
Crooke, Colin 138
Dennis, Roy 1, 2, 6, 9, 108, 132, 167, 187, 203
Duffy, Kevin 183
Ellis, Pete 78
Evans, Rhian 200, 203, 209
Folkestad, Alv Ottar 203, 208
Green Rhys 182, 183
Harvie-Brown, John Alexander 87, 90
Helander, Bjørn 195
Langslow, Derek 136
Macdonald, Archie 190–1
Macdonald, Duncan 87–8
Macgillivray, William 33, 39, 218
Mackenzie, Alexander 35
MacLennan, Alison 194, 198
Macpherson, Rev. H. A. 87
Madders, Mike 166, 168, 207
Marquiss, Mick 178
McGowan, Bob 92
Mee, Dr Allan 199, 201
Miles, Hugh 134, 154
Miller Mundy, Andy 33, 163, 167
Misund, Harald 20, 37, 49–50, 119, 122–4, 138, 184, 188
Morton, Keith 168
Mudge, Greg 184
Nellist, Kate 167, 212–13
Newton, Dr Ian 107
Nye, Pete 136
Oehme, Gunther 20
Pienkowski, Mike 182
Rae, Stuart 179
Ratcliffe, Dr Derek 107
Rooke, Flight Lieutenant Steve 137–8, 185
Sandeman, Pat 1, 3, 4
Scott, Carol 132
Sexton, Dave 166, 168, 196–7, 210
Simison, Ronnie 59
Smith, Claire 203, 204, 209
Smith, Ken 179
Thompson, Kate 183
Thorne, Debbie 210
Waterston, George 1, 6
Watt, George 132
Weir, Tom 4
Willgohs, Johann 5, 8, 20, 110, 138
Wolley, John 34–5, 73, 92